Venture Labor

Acting with Technology

Bonnie Nardi, Victor Kaptelinin, and Kirsten Foot, editors

Venture Labor

Work and the Burden of Risk in Innovative Industries

Gina Neff

The MIT Press
Cambridge, Massachusetts
London, England

MIT Press books may be purchased at special quantity discounts for business or sales promotional use. For information, please email special_sales@mitpress.mit.edu or write to Special Sales Department, The MIT Press, 55 Hayward Street, Cambridge, MA 02142.

This book was set in Stone Sans and Stone Serif on 9/13 pt by Toppan Best-set Premedia Limited, Hong Kong. Printed and bound in the United States of America.

Library of Congress Cataloging-in-Publication Data

Neff, Gina, 1971–
Venture labor : work and the burden of risk in innovative industries / Gina Neff.
    p.   cm.—(Acting with technology)
Includes bibliographical references and index.
ISBN 978-0-262-01748-0 (hardcover : alk. paper)
1. High technology industries—United States—Employees. 2. High technology industries—United States—Management. 3. Technological innovations—Economic aspects—United States. I. Title.
HD8039.H542U654   2012
331.700973—dc23
                                                                2011039214

10 9 8 7 6 5 4 3 2 1

For Philip Howard, eleven times more

Contents

Preface

One of my mentors, Herbert Gans, warned me that I should get this book out quickly before people forget that the Internet boom ever happened. Yet optimism about our digital future is back in vogue, years after the first dot-com bust. Indeed, if people have forgotten that the Internet boom happened, it is because new Web 2.0, cloud computing, social media, and crowdsourcing have become the exciting new digital tools and social trends. But in an important way, the backstory for the industries behind these innovations has not changed since the Internet boom. While the start-up companies on this current wave of investment are not creating webpages or "B2B" e-commerce sites, they are relying on many of the same phenomena as the companies that were started in the lofts and apartments of New York's Silicon Alley in the 1990s. Innovation, then and now, relies on the phenomena I call *venture labor*, the investment of "everyday entrepreneurs" in the companies where they work.

With the present euphoria around social media and Web 2.0, we can see how easily we can forget, though. Many of the lessons from the beginnings of the Internet, especially how small companies created news, information, and entertainment for digital media, are relevant for understanding the directions that today's companies will take and how our media will look in the future. The Internet may no longer be in its infancy, but it is still far from maturation and many innovations and ideas will continue to be developed.

Although *venture labor* as everyday entrepreneurship gained visibility and became manifest during the dot-com boom of the late 1990s, the phenomena both predates and extends the Web 1.0 era. As a new wave of social media and new media companies hold initial public offerings of stock, we see yet another cycle of people investing in the companies where they work, getting in "on the ground floor" of creative and technical opportunities in a sector that appears to be expanding.

This book lays out the motivations for why and how people work for these types of ventures. Start-up companies and emerging media will always depend on people who work to turn innovative ideas into reality. Rescuing the historical lessons of the first wave of Internet companies can help the next wave understand how to better support the everyday entrepreneurs that these companies rely on. The media legacy of Silicon Alley lives on, and the process of venture labor is very much alive.

Acknowledgments

One of the key lessons of this book is that entrepreneurial projects always rely on collective efforts. This book is no exception.

I am exceedingly grateful to the people who opened their lives up to me. Their stories inspired this project, and I hope that the final book reflects the deep respect that I feel for their work and their accounts of it. Chris McCarthy was the first to introduce me to the (then) strange new world of Silicon Alley, and without him I could have never started this project. Courtney Pulitzer also graciously provided me with years of her archives and materials about Silicon Alley, and without her work the social network maps of Silicon Alley wouldn't exist.

My mentors shaped the project from the beginning—Sharon Zukin, Charles Kadushin, and Charles Smith at the Graduate Center of the City University of New York and Harrison White and Herb Gans at Columbia. I'm deeply indebted to David Stark and Monique Girard—both of whom pushed me to develop this project into the present book. The community that David and Monique created through the graduate fellows and affiliate faculty of Columbia's Center for Organizational Innovation was indispensible to me. Over many dinners and discussions, these friends and colleagues patiently listened to and passionately debated the ideas that ended up here.

Colleagues in a wide variety of disciplines at other institutions greatly contributed to this project. While I can't possibly name them all, I would like to thank David Kirsch, Fred Turner, and Pablo Boczkowski in particular. Several people gave me extremely helpful comments that helped improve the final version of this book (and prevented some embarrassing blunders from finding their way into print), including Dan Cornfield, Howard Aldrich, and the anonymous peer reviewers. The book greatly benefited from talks and workshops at the University of Wisconsin-Madison, the University of California, San Diego, Indiana University, the New School, the Annenberg School for Communication at the University of Pennsylvania, St. Mary's University in Halifax, and the University of Maryland's Webshop, which supported a whole generation of young scholars of the Internet, creating lasting social bonds among us.

I received valuable support for this project from the University of California's Institute for Labor and Employment and its director Ruth Milkman, the Helen Riaboff Whiteley Center at Friday Harbor Laboratories, and the University of Washington Department of Communication. The Center for Advanced Study in the Behavioral Sciences at Stanford provided an office and a welcomed intellectual community. The fine people of Herkimer Coffee, especially the incomparable Kara MacDonald, provided treasured space and stimulation for completing this book.

Margy Avery at the MIT Press has been incredibly supportive of this project. Katie Persons has been wonderful in seeing this book through the editorial process, and Kathy Caruso marshaled my unwieldy prose into line with her keen editing and fought a hurricane to do so on time. The Acting with Technology series editors, Bonnie Nardi, Victor Kaptelinin, and Kirsten Foot, provided the crucial feedback that turned a manuscript into a book.

One of my main sources of social support in graduate school was a group we formed for studying for our first-year exams. We have since stuck together through weddings, births, dissertation anxiety, graduation, and tenure. These amazing women supported this project, and much more important its author, from the very beginning and in the darkest moments. For that I would like to thank Ariel Ducey, Lorna Mason, Ananya Mukherjea, and Michelle Ronda. As a new faculty member at the University of Washington, I became part of a multidisciplinary tenure support group, an amazing group of "powerettes" that includes Cindy Perry, Leigh Mercer, Julie Bierer, Barbara Citko, Helene Starks, and, in particular, my collaborator and co-conspirator, Carrie Sturts Dossick. Sue Gaylard continues to be an amazing friend and writing buddy and makes working in Suzzallo Library something to look forward to.

I couldn't have completed this book without the friendship and support of my dearest friends, Crickett Rumley, Jenn Lena, Betsy Wissinger, and Joby Dixon. I owe so much to my parents, Gene Neff and Susan Neff. Philip Howard served double duty as my husband and my colleague, cheering me on, reviewing drafts, and making the time for me the time to finish this book. For that, I'd like to thank him and my beautiful sons Hammer and Gordon for indulging me on my early-morning café editing sessions, and for teaching me that writing a book is even more difficult and time-consuming than having a baby or two.

Any glimmers of brilliance in the manuscript are the result of the constructive criticism of all of these people. The author bears the burden of risk for any remaining flaws.

Salt Spring Island, British Columbia

1 The Social Risks of the Dot-Com Era

"How could we have been so stupid?"

Sophie had been a heroine, of sorts, during the rapid growth of the Internet industry.[1] She joined the great dot-com boom of the 1990s and worked her way up from an administrative assistant with an advertising agency to a project manager for a web design firm. Sophie and her husband moved from New York in 1998, a time when the money was good and growth was steady, to Silicon Valley in California where the dizzying, frenzied pace of start-up activity held the promise of fantastic money coupled with phenomenal potential for business growth.

During my visit with her in the summer of 2002, Sophie questioned not just that move to the West Coast, but all of her career choices. She felt relatively lucky: After several months of unemployment for both her and her husband, Sophie had found work with a large nonprofit group that would utilize her skills in managing online projects. At the time of my visit, her husband was still unemployed. To Sophie, the confidence that led them both to stake their futures to a booming industry and to transfer their careers from New York's Internet industry to Silicon Valley now seemed like the hubris of naïve youth.

Four years earlier, I didn't think Sophie was stupid at all: rather, I thought I was the one making a bad economic choice by going to graduate school at the very moment many of my friends were taking interesting, creative jobs in New York's emerging Internet industry. I was jealous—the same friends who shared horror stories with me of underpaid, dead-end, and temporary administrative assistant jobs after college were suddenly excited about their work and their careers. In doing research in their field, I had the opportunity to do the kind of "deep hanging out,"[2] as the anthropologist Clifford Geertz has called it, with people who were doing things that I thought I could have been doing too had I not made the seeming financially stupid decision to go to graduate school. My friends—just like

the people I interviewed for this book—were doing interesting, intellectu-
ally challenging work, creating cutting-edge content for a new medium,
and building a like-minded community engaged in nothing less than
changing the way society communicates. At dot-com launch parties, gatherings for drinks in hipster East Village bars, and artsy office lofts, young
people like Sophie seemed to have made the smart but easy choice.

In their classic handbook of field method, Schatzman and Strauss advise
the field researcher to be "particularly sensitive to his own interpreted
experience," for even "the most subtle of his own surprises . . . is a sign
that some expectation or hypothesis of his has been altered or even shat-
tered."[3] My own shattering moment was when my jealousy of dot-com
highfliers turned into sympathy for Sophie and others for whom years of
hard work seemed to simply evaporate. Neither greed nor hubris nor stu-
pidity were the reason that Sophie and thousands like her found themselves out of work in an industry that had collapsed. But Sophie felt, as
many people within advanced industrialized Western economies do to a
greater or lesser degree, that she alone was responsible for her economic
success or failure. Many people within the nascent Internet industry of the
mid- 1990s and early 2000s felt assured that their unique combination of
skills, business contacts and connections, and experience in a rapidly
growing, new high-tech field could protect them from any turbulence the
emerging industry. One job in the industry would lead to another; friends
could always be counted on for leads to new work; the industry as a whole
would always expand. As Sophie so eloquently phrased this confidence,
"We thought the risk was that our company wouldn't go IPO, or maybe
fail. We never thought *all* the companies would fail."

While Sophie's statement exaggerated somewhat the extent of the
downsizing in the Internet industry, the bursting of the dot-com bubble
did affect employees' abilities to find other employment. One cover of the
Silicon Alley Reporter, a New York Internet industry trade magazine, illus-
trated the "year of the dot-com crash" as a burning Zeppelin, a technological advance that represented great optimism at its inception and grand
folly at the time of its crash.[4] Like the Zeppelin, working in the high-flying
dot-coms seemed like a bad idea only after the crash.

Many of the people I interviewed for this book articulated an understanding of the economic uncertainty of the Internet industry that confirms what social theorist Ulrich Beck predicted—economic risk in modern
life has become increasingly privatized and individualized.[5] But economic
risk is a social, collective phenomenon and depends on factors that are
beyond any single person's control, even if the repercussions of those risks

are privatized. In this way the economic risks (and indeed most risks) we face have a social component. Will the demand for our skills and labor increase in the future? Will the economy support our jobs? Will our companies be successful and continue to need our labor, or will they fail, shift jobs overseas, or subcontract out our services? People must answer questions like these to navigate the economy now, and such questions reflect how the chances we take depend on social factors, even if we feel we are solely responsible for the repercussions.

Risk Is Social

In retrospect, the small dot-com companies based in New York in the mid-1990s until the 2000 crash seemed financially risky. We may ask ourselves retrospectively, like Sophie did, how anyone could have thought that so many young start-up companies would survive much less potentially morph into the next Microsoft or Google. We could ask ourselves *why* Sophie, her husband, and people like them staked their hopes on the Internet industry, or we could try to prove their logic and economic reasoning right or wrong. However, the better story—and the more powerful social and cultural explanation—lies in understanding the social processes that made the risks that they took seem safe, natural, and routine. Rather than ask why dot-commers made those so-called stupid choices, we should instead ask what made taking such chances seem like a good idea at the time, and examine the economic and social processes and cultural contexts for those decisions.

In this book I argue that a changing cultural and political landscape created the context for the dot-com boom to happen. New cultural attitudes toward risk—attitudes that predated the Internet boom but found a newly entrepreneurial subject before and during the dot-com boom—fostered the euphoria around the industry and attracted workers to new companies. Magazine and newspaper articles celebrated and naturalized these economic risks by urging a casual, even positive, attitude toward losing one's job. Such portrayals framed economic and financial risks as inevitable, necessary, and beneficial for one's career and companies, reinforcing cultural messages about the attractiveness of risk. As phrased at the time by cultural historian Jackson Lears, "We're seeing a resurgence of risk both as a necessity of economic success and as a mark of what is fashionable. . . . What's really happening is risk is now cool."[6]

And nowhere was this more visible than among the young, urban, educated workforce of the first wave of dot-coms. Any downside to such risks was rhetorically presented in the business and popular press and in the discourses that circulated among workers themselves as a narrow spectrum of possibility, which could be easily overcome, managed, or ignored.

This framing, I argue, is the way that economic risks are socially constructed. Whether people talk about risks with each other or read about risks in the news media, the frames they use influence what they and others perceive as risky and how responses are constructed to those risks. As Mary Douglas and Aaron Wildavsky have written, it is less important to focus on the "real risks out there" than it is to study these processes that make things appear risky or not.[7] Certainly, at one point in the dot-com era, working for a start-up did not appear as risky as it seemed to be after the dot-com crash. What I propose is that we shift the lens of our collective analysis away from how individuals make decisions toward the *cultural frames* for those decisions. Rather than thinking of the economists' categories of risk-loving, risk-neutral, or risk-adverse, thinking of risk as work allows us to recast attachments and affiliations to risk. People can be adept at this work, enjoy it, or dislike it. In other words, people, organizations, and institutions can frame uncertainty in different ways and to different ends. But taken together, this framing of risk has very real consequences.

Within economic literature there is a clear distinction between risk and uncertainty. In Frank Knight's classic formulation, risks are knowable, calculable, and probabilistic. In this sense, someone can hedge against risk and take a risk with knowledge of the possible outcomes, regardless of whether the risk pays out or not. *Uncertainty*, in Knight's theory, is unknowable, like the odds of the complete collapse of the economy. Risk, in Knight's definition, entails careful choice among several options, balancing risks with potential rewards, and balancing safer and riskier options. Knight argued that real entrepreneurial profits do not come from managing risks because risks were known, which makes it difficult for one entrepreneur to have an advantage over others or over the market. Rather, profit comes from exploiting uncertainties and managing the incalculable.[8] The strategies for managing risk that may work for an investor or in a stock portfolio won't necessarily work for an employee. Narratives and discourses about the market during the dot-com boom made individual risk acceptable and framed risk as cool. These narratives were reflected in the ways in which people talked after the dot-com crash about the risks they took.

One of the reasons new companies frame uncertainty in particular ways is because they need people willing to take risks. The discourses of risk during the dot-com boom encouraged people to take risks and not to fear failure. Two reporters suggested that the reason for this change in attitudes was "to encourage the kind of risk-taking needed to spur entrepreneurs to hack away repeatedly at the American Dream."[9] Discourses expressed this symbiotic relationship between the risks that companies and their investors take and the risks borne by labor. Tom Perkins a partner of Kleiner Perkins Caufield & Byers, one of Silicon Valley's top venture capital firms, said, "It is to a certain extent in the best interest of venture capitalists to encourage people to keep trying, to not be afraid of failure. We need people to take a chance."[10]Another venture capitalist and new economy guru Esther Dyson also lamented in a *New York Times Magazine* article in 1998 that ideas were cheap, but people to work on those ideas were not; she claimed that the problem with the U.S. technology industry was that there were too many ideas and not enough talent to implement them. As she phrased it, "The best idea in the world won't go anywhere without someone to carry it through."[11] This book examines this discourse surrounding a changing culture of risk that encouraged people to work in the dot-com industry. What emerged with the rise of dot-coms was a powerful cultural message that workers should willingly take risks as the requirements for their jobs. This book examines these discourses about risk and their implications for job security.

Sociologists understand that even most private and personal of decisions are structured by larger social forces. These "collective forces," Emile Durkheim wrote, "determine our behavior from without."[12] The decisions that Sophie and her husband and legions of other young people made to join what appeared to be—and was often called—a dot-com gold rush were also shaped and informed by collective forces as well. Researchers often examine economic and financial decisions using analytic tools that focus on individual behaviors rather than these social or structural forces, and this is especially true for research on risk taking. Studying how individuals make decisions has been an important part of both economics and sociology, but the collective forces that shape individuals' frameworks for decision making and risk taking—or the social structures that shape *how* people make decisions—are important to understand when it comes to questions concerning economic behavior. Such studies also show the role for media and communication processes in shaping the economy. Behavioral theories of risk have well-developed models of individual decision making, but there are fewer studies that connect individual perception of risk to the social and institutional forces that influence those perceptions. Understanding how risk works in ordinary economic life is not an easy task:

misperceptions linger in what Fred Block called the "subterranean level of economic understanding,"[13] misperceptions that rational actors take risks in exchange for possible rewards and that profits are the payoffs for risks. This is the challenge set out for this book.

While I do not describe people in terms of rational action, I do take seriously how they talk and frame their decisions as evidence of their discursive work with risk. I do not intend to argue in this book that their decisions were good or bad, rational or irrational. My research is predicated on the assumption that we can take seriously the actions and discourses of contemporary subjects without having to resort to the false dichotomous poles of framing them either as fully formed rational actors with awareness of and power over their lives or as subjects duped by capitalist relations into a lulled state of false consciousness. The people I interviewed and observed managed their risk through talk, and the ways they articulated their positions within the economy helped them make sense of their choices. This process of articulation reflects, perhaps imperfectly, how they represented risk in their lives to themselves and to others, and such representations had and continue to have real impact in the economy.

Work Is Riskier

I argue that the dot-com boom occurred at a moment of transition in U.S. economic history toward riskier work, and the entrepreneurial spirit that people enacted during the boom was a response to this economic transition. In technology industries, employee risk taking has perhaps been the most visible—and the most attractive—as stock-option millionaires were created by the lucrative initial public offerings, but there are many other ways that economic risk has increased at work. Jobs are less secure, with lifetime employment a thing of the past. Layoffs, now a commonplace phenomenon, were relatively unheard of in the corporate world before the 1970s.[14] For those working in the Internet industry during the dot-com boom, taking risks was seen as the best among otherwise limited options within the economy.

An array of types of support from families, jobs, and governments have historically helped people mitigate the economic risks they face. Imagine, if you faced a period of unemployment, what resources and upon whom could you rely? If, like many within the Internet industry, the booming, growing industry you worked in suddenly began shedding jobs, what resources would you have to support an extended layoff or retooling? Within the United States, you might be able to count on months of unemployment benefits, depending on whether you had worked as a permanent employee or a contractor. Your family and friends form another kind of support, and your workplace may provide some support as well. As Jacob Hacker argues, workplace and government supports are not as robust as they once were and are not adequate for the risks people now face. This "risk privatization" means that social supports for risks "cover a declining portion of the salient risks faced by citizens," and as a result "many of the most potent threats to income are increasingly faced by families and individuals on their own, rather than by collective intermediaries."[15] These risks that people face at work are yet another example of how risk is socially structured and determined, yet privately handled and managed. Economic downturns, company layoffs, booms and busts—these are collective phenomena, but people attribute managing these risks to individual pluck.

Three economic forces increased the level of economic risks people bore in the late twentieth and early twenty-first centuries: the increasing "financialization" of the American economy; rapidly changing valuations of work, products, and services within the new economy; and the widespread diffusion of flexible work practices. These processes, outlined here, connect macroeconomic changes with risks managed and experienced on the individual level.

The first of these processes is growth of financialization, or the emphasis that companies place on profits from financial capital over profits from productive capital.[16] While a new body of research has begun much-needed examinations of financial markets as social phenomena, there is still little analysis of how this increasing role for finance within the economy influences decisions and lives *outside* finance.[17] The Internet industry in the late 1990s was one of the most apparent fields in which to study the volatile combination of the power of speculative financial capital and increasing employment risk. The current "global financial crisis" that began with the burst of the U.S. housing bubble in 2008 is yet another example of the devastating impact that financial discourses can have for markets, jobs, and consumers.

The second process is valuation, or the social construction of negotiation around economic value. Within the rapidly expanding Internet industry, uncertainty about the future directions of the industry led to shifting valuations and evaluation about the value of work, products, and services. Economic geographer Nigel Thrift argued that these valuations do meaningful work in the service of capitalism.[18] Moments of open or conflicting valuations can create opportunities for a form of arbitrage—for betting that the value of a good, a service, or a skill will increase in value when transformed by a different valuation schema. This is, in part, what Harrison White refers to when he claims that "business activities are sustained . . . only as common discourses are generated and shared in common histories."[19]

The third process is the rise of flexible work practices. Many scholars have studied the changing relationship between individuals and work, noting that as jobs in Western economies shifted from manufacturing to services an employment culture of "flexibility," instability and insecurity emerged.[20] Patterns of employment since the late 1970s have favored more "flexible" forms of work, the increased globalization of trade and finance, and what Manuel Castells calls the "informationalization of work."[21] These changes have had repercussions for workers. For example, sociologist Vicki Smith argues that employees now see risk taking as so inherent to their jobs that many of them see risk as the only access to economic opportunities: "When corporations no longer buffer their workers from the uncertainty of production and employment workers must take risks and expand great personal and group resources to control that uncertainty themselves."[22] In other words, this new ideology of flexibility links the lack of job security to opportunity, making risky work attractive.

Smith characterized this new work environment as one in which "employment instability, decentered control, and work intensification run across the occupational spectrum."[23] What this means is professionals can no longer expect long-term, secure jobs. The increase in contract and temporary work increases companies' options and flexibility by distributing part of the burden of risk to external subcontractors and self-employed freelancers. When times are good there is work, and in downturns contracts are not renewed. This is an explicit *externalization* of costs by a firm onto workers. There is widespread agreement that these changes are the result of increasingly complex financial and economic interdependencies. As a result, new industrial relations of uncertainty are emerging. What scholars have called the "Fordist" social arrangement of production—with its reliance on both stable employment relations and government-backed social safety nets—have been replaced or weakened in many industries, especially in what has been called the new economy. Flexibility along with weakened social protections mean that postindustrial work depends upon workers' ability to manage uncertainty in unprecedented ways, as we'll see in the chapters that follow. Rosabeth Moss Kanter has called this a move from employment security to employability security. She argues that the high-tech industries provide an unfortunate model of this practice for the rest of the economy: "Instead of counting on long-term employment with a single firm, they increasingly depend on their employability by many firms. The shift from employment security to employability security implies a fundamental change in what people should expect from their employers—and how employers should think about their interests and obligations."[24]

Understanding those decisions in context requires an analysis of the meanings and symbols that actors attach to that risk. Studies of innovation show that actors' values toward risk taking in financial, technological, and market realms can vary across institutional environments, creating regional areas conducive for innovation or clusters of creative people that fuel urban growth. Levels of entrepreneurial activity can vary across cultures, explained by institutional factors such as patent supports, intellectual property arrangements, and markets. Cultural differences in the approach to risk can also influence how people act.

Ulrich Beck and those who have extended his *Risk Society* thesis question the *social* characteristics of economic risk. They argue that in light of a weakening social safety net, individuals are generally being forced to bear more risk, and thus economic life is riskier now than in previous eras. This individualization of risk means a greater exposure on a macro-level to environmental risks, to job-related risks, and, especially in light of decreasing social supports, to the risks that accompany business cycles. Beck argues that the ability to bear risk can be stratified—that is, some people and institutions are better equipped to manage risk than others, and this stratification works with and amplifies already existing stratification regimes.[25]

Beck primarily deals with the shift of risk from economic organizations—corporations, unions, governments—onto individuals, and, in this way, implies that these organizations could distribute risks more fairly if the political will to do so existed. For Beck, risk equals a kind of danger that workers now face alone—without the social and economic protections previously afforded by organizations and institutions.

Beck argues that people now see many kinds of risks as their individual responsibility, and indeed this seems to be the case. For Beck, the individualization of risk is reflected in the economy through a shrinking social safety net and the increased exposure of employees to market forces and in the environment through a pervasive exposure to environmental "bads" such as toxins and pollution. Beck predicted the rise of a *risk society* in reaction to the pervasiveness of risk, especially environmental ones. While knowledge about risks and the power to do something may not be equally distributed, the pervasiveness of risks, Beck argued, would lead to increased collective action to prevent and manage risks. But, just as Marx never predicted that capitalism could produce such a large and strong middle class, Beck, writing before the economic downturn that began in 2000, did not foresee how ideologies of economic risk would continue to make risk seem attractive, at least among the U.S. labor force.

Framing Risk

This book examines how people frame the risks of their jobs. Framing is an important communication concept because it shows how the contested terrain over concepts and ideas shapes how people, in turn, act. The chances that people took during the dot-com boom were not merely the result of the poor or wise decisions of many individuals involved. Nor were they simply rational trade-offs between risk and rewards. Social processes and collective forces structured how people perceived and took risks, demarcating the range of available choices and the amount of risk apparent to themselves in their choices.

I argue that these social processes in part helped create the dot-com boom. People's desire and need to take economic risks stemmed from a *lack* of job security and an increase in employment flexibility—not the other way around. Because work in general became riskier, people became more willing to take more risks. The discursive, communicative functions of the new economy helped construct this new economic reality. The dot-com boom created a vicious cycle—taking risks seemed to be the only way to get ahead—encouraging entrepreneurial behavior from people in the industry, which in turn signaled to others that taking risks was a good idea. These processes are not neutral since political and economic power plays a substantial role in determining which discursive frames are important, how they function within the economy, and for whom they generate profit. In the chapters that follow, I explain how social forces *naturalize* economic risks. I show how postindustrial workplaces integrate discourses of economic risks into their businesses by encouraging employees to "buy-into" their companies' goals and to invest their time, energy, passion, and money. Within this historical context, some employees act like financial investors investing their labor in exchange for potential entrepreneurial rewards and business risks.

Entrepreneurial Workers Not Entrepreneurs

This book focuses on the rise of entrepreneurial behavior among people who worked in emerging technology industries in the late 1990s. Their behavior was not independent from larger economic trends, nor was it solely the result of the rapid pace of technological change under way. Specifically, their behavior was in part a product of a particular historical moment in which economic risk shifted away from collective responsibility toward individual responsibility. It is what Jacob Hacker calls the "great risk shift,"[26] or a societal move toward greater personal responsibility for economic and financial well-being and away from responsibility shared within companies and by the nation-state. One headline in 2000 summarized the shift work was undergoing in the new economy culture as follows: "Risk and reward are key, not job loyalty."[27] These macrostructural changes in the contemporary American economy—and the cultural shifts around these economic changes—shaped the rise of the dot-com era, not the other way around. As John Child and Rita McGrath say in their article on "unfettered" organizations in the information age, "When a society's organizations thrust a large number of its citizens into a condition of permanent survival-oriented tension, it would be remarkable indeed if the effects were benign."[28]

The dot-com era was a response to these changes, not the cause of them. The social shift toward increased employment flexibility created a fertile landscape for entrepreneurship and risk taking. Having high skills that were in demand, people working in technology attempted to find their own answer to uncertainty in the U.S. economy by taking risks and chances that they at least felt some modicum of control over. High-tech companies seemed to welcome and encourage employees' risk taking as contributing to a more democratic and participatory form of organization, even within a longer trend that devastated a culture of corporate loyalty to employees. Part of this is attributable to the rise of "market populism" with the stock market boom, as Thomas Frank argued,[29] but is, I argue also, part of a larger shift within the U.S. economy. The dot-com boom helped glorify risks— and shifted social and economic uncertainties to individually accounted risks. People accepted and welcomed risk because taking risks offered a semblance of choice in an era when many things were out of ordinary employees' control.

The notion that risk taking will ultimately be rewarded is a deeply held value within American business culture. Taking risks at work is inexorably linked to the promise of possible wealth, so much so that company founders' stories commonly highlight what economic risks a founder faced and how that founder outsmarted the market. Economists have long viewed profits as the returns for taking risks, but as an *ideology* risk taking provides a rationalization for economic stratification that is almost as powerful and complete as the idea of meritocracy: those who take risks get ahead, and those who don't are left behind. The rhetoric of the new economy echoed this adage as if it were an established fact.

This book examines how people, as active social agents, navigated and adapted to these dramatic social changes and the ways in which they attempted to control and manage these risks. Interviews I did for this book show that people working within the Internet industry during the dot-com boom and crash understood the economic risks they faced in *very individual terms*. That is, they thought they understood all the risks they faced, and they thought that they had the power to hedge against or profit from these risks independently of what was happening within the economy at large. With the dot-com crash that began when stock prices tumbled in early 2000, many people abruptly realized how little power they actually wielded over larger economic forces. The economic risks that they took may have been social, but the responses to the outcomes of those risks were experienced personally.

Political scientist Jacob Hacker has argued in *The Great Risk Shift* that shifting risk away from government has meant "risk is on the workers, not shared between companies and workers."[30] As he observes, the point is not whether they do better or worse in this system, "but that they face far more uncertainty and risk."[31] That is, there is more volatility in the incomes and lives of the American workforce than any time in the post-World War II era. Hacker writes that it is a mystery why corporate and political leaders lag in responding to the increased insecurity that Americans face.[32] The real mystery to me, though, is why have people been so willing to accept this risk? Some of the answer is in how risk was made attractive during the dot-com era as something to be embraced, rather than feared. As Joost Van Loon has argued, the discussions of risk in political talk and media discourse are *autopoetic*, which means that as the message is communicated it reproduces itself. In other words, the more risk is discussed, the more it remakes a riskier society. It is in this way, Van Loon contends, that the "social order is gradually being eroded at the cultural level."[33]

In addition to the political and economic changes that brought about the new economy, there were new ways of talking about risk. These new ways of discussing risk led to new ways in which people managed, dealt with, and expressed risk. And, as we'll see in this chapter, these ways of dealing with risk in turn led to new exposures. The dot-com era was both a rhetorical and discursive strategy for reframing work relationships of the new economy. As Nigel Thrift put it, it was in part the "the romance" of exciting, interesting, and risky new jobs in a rapidly growing field, not the financial rewards, that attracted many people to work in the dot-com boom.[34] This allure combined with the newness of many companies in the Internet industry meant that workplace-level changes flourished in the industry. With few established conventions or work practices, dot-coms could reinvent professional work as fun, young, and exciting, turning jobs from white-collar into what Andrew Ross has called "no-collar."[35] As new companies in a new industry, they emerged distinct from established institutions and models of work in older industries, and the individualism of this new era could thrive. Fred Turner has convincingly argued that early cyberculture was deeply influenced by a particular brand of political libertarianism of the countercultural movement.[36] Similarly, the *culture* of risk was just as important to the rise of the commercial Internet. Surely, the economic rhetoric of both Clinton and Reagan resonated with the changes that people felt; in turn, political talk helped shape a culture that accepted economic risk. While wide-reaching economic and political changes precede the dot-com boom, these changes informed and supported a cultural shift that occurred—making risk a central narrative in how people talk about their work. Not only did these changes set the ground for a new economy, but the discursive shift toward risky work also helped constitute new work practices and new relations to work. Decades of industrial change hinged on a fulcrum between two regimes—an "old" industrial economy in which economic growth and cultural norms supported stable employment practices in a "new" information, cultural, and technological economy driven by highly individualized and flexible work. This distinction between the old and new economy I have shown in this chapter is not a clear-cut one, and the rhetoric and reality of the scope of the economic changes was hotly debated at the time. But accompanying this discussion of a new, "renewed" innovative economy was talk of risk taking, entrepreneurship, and individual initiative that informed the ways in which people think about their jobs. It is in this economic, political, and cultural environment that a new medium arose, and embedded into these new technologies and messages were the entrepreneurial values of the people who were creating it, as we will begin to see in the next chapter.

Economic and financial risk is not something necessarily natural and innate, but rather constructed in part from the discourses that surround it. As Louise Amoore has argued, how risk is framed concretely shapes social practices. Management consultants influenced employees' perceptions of the risks of globalization and the actions that the employees took as a consequence of these perceptions. If the risks of globalization seem inevitable, Amoore argues, then there exists a very different range of choices that people feel they can make.[37] While Amoore focused on discussions of globalization, the same holds true for how workers welcomed risk, trumping any discussion of uncertainty or structural change in the U.S. labor market.

In a similar way, discussion of risk in the new economy put a positive spin on the discourse of unforeseen events, one in which people learned to accept that they could profit from uncertainty and should embrace—not fear—the corporate changes under way. Just as managerial doctrines about globalization encouraged employees to be more entrepreneurial, the discourse of risk served a powerful symbolic function to get employees to accept more uncertainty within their jobs and within the economy. The lure of risk—the potential for payout—adds an element of choice, that people are choosing to accept risk rather than merely accepting the consequences of economic structural change.

Geoffrey Hodgson, in his critique of what British scholars call "the learning economy," argues that at the root of this utopian vision of market individualism is the notion that the individual is the best judge of his or her own needs. However, this vision of society ignores the fact that individuals are socially formed, through both the process of socialization and people's knowledge of possible choices.[38] As a consequence, economic decisions are more complex to disentangle: do people make choices based on their best rational judgments or because of how they interpret the multiple competing signals? As Hodgson writes, a connected set of developments of increasing economic complexity and more advanced knowledge "make the distinction in practice between employment and self-employment all the more difficult to uphold."[39] As a social endeavor, work itself has been largely individualized, Hodgson argues: "The employment contract is in large measure a convenient fiction, couched in the individualistic categories of modern contract law, which in fact masks the social and co-operative character of all productive activity."[40]

The people who study governmentality have shown how we internalize these economic desires into our being. No longer are work relations constructed as conflictual but rather framed as a way that we *want* to be working. The culture of the era meant people internalized the message that taking risks was desirable and, in turn, internalized the lack of economic security within their jobs.

There is an increased ability among people to identify these risks on a macroeconomic level. Our economic data is more finely calibrated than ever, and there is more widespread knowledge about the economy with the increase in media reporting about business and the stock market. Political scientist Mark Smith found that there has been a radical shift in the way the press has talked about the economy. The amount of front-page reporting of economics in the *New York Times* doubled from 1973 to 2004 compared to the previous twenty-year period. In a public opinion survey, only 17 percent of those surveyed from 1946 to 1972 mentioned the economy as one of the most pressing problems; from 1973 to 2006, a shocking 73 percent mentioned the economy.[41]

However, this knowledge is now matched with distrust in the institutions that could help individuals hedge their bets. As Van Loon put it, in a risk society "closures offered by expertise, legislation and moral panics are no longer met with trust in the systems that produced them."[42] We have replaced the old economy with something we no longer trust, although we have more information about the risks. Less trust in institutions such as the government and corporations means people are placing relatively more trust in themselves, whether or not by necessity. This in turn further diminishes the ways in which people press institutions like the government to provide security. To a large degree, the arrangements that used to buffer American workers from economic insecurity—such as expectations of corporate job loyalty and increased government support for jobs growth—have been replaced by mechanisms that place the burden of risk more squarely on the shoulders of individuals. How have people adapted to these changes through their decisions, their career choices, and their approach to their jobs?

Framing economic risks as desirable is one of the enduring social consequences of the dot-com era. After decades of growing concern over job security, downsizing, and corporate cutbacks in the United States, the lure of risk during the dot-com boom made the lack of job security seem like a choice—and a pretty good one at that—for people working in the potentially lucrative technology industry. The lure of risk—and by this I mean the idea of *taking* chances—replaced the fear of uncertainty as the predominant economic rhetoric during the Internet boom. This shift is subtle but important as *risk* and *risk taking* in economic life now imply active choices while *uncertainty* connotes economic passivity and forces beyond one's own control. For high-tech firms and start-up Internet companies, skyrocketing stock prices during the late 1990s gave risk the shiny luster of potential wealth for all employees, regardless of the chances for payoffs. Risk provided a justification in individual terms for both the profits and losses that came with the stock market crash in 2000. When the crash happened, the people I interviewed questioned what was wrong with *themselves* rather than what was wrong with the economy. Within a larger social and political context of flexible work, risk-taking privileges potential individual rewards and losses over collective responsibility and group, organizational, or structural causes of either prosperity or poverty. In *The Disposable American*, Louis Uchitelle argued that even when we know layoffs are not the fault of workers, people who are laid off experience them as personal failures.[43] American society takes actions that stem from collective social and economic forces and turns them into something private and personal. In other words, society has individualized the outcomes of collective economic risk.

Risks and rewards at work are not new, of course, but changes in the organization of work mean that individuals now bear most of the costs of flexibility and are responsible for activities previously thought of as within the purview of companies. Individual workers are less protected by their companies from the economic risks that companies face, and, with the fraying of the social safety nets of union protection and government support, workers are facing these risks alone. It is as if the logic of American capitalism replaced the metaphor of "climbing the ladder" for professional work with that of jumping aboard a ship that has yet to come in. In the words of a web site producer, "it is really up to you to manage" the risks of a market downturn, of losing a job, of becoming technologically obsolete. The lure of the potential payouts for these kinds of jobs made taking risks more attractive while masking the downsides of risk and insecurity.

Defining Venture Labor

One strategy for managing the risk of contemporary work is what I call *venture labor*. Venture labor is the investment of time, energy, human capital, and other personal resources that ordinary employees make in the companies where they work. Venture labor is the explicit expression of entrepreneurial values by nonentrepreneurs. Venture labor refers to an investment by employees into their companies or how they talk about their time at work as an investment. When people think of their jobs as an investment or as having a future payoff other than regular wages, they embody venture labor. Venture labor is the way in which people act like entrepreneurs and bear some of the risks of their companies. Venture labor includes the entrepreneurial aspects of work—how people behave as if they have *ownership* in their companies, even when they are not actual owners.

Venture labor is about people taking risks for their jobs, as much as it is about their subjective embrace of that risk.

Venture labor is one way that employees adapt to bearing the economic and financial risks of the companies for which they work. The term is a play on venture capital, the private investment on which new companies rely. People, venture laborers, can invest in their companies in many different ways. Employee retirement funds may be invested in company stock, and in new companies there are often tacit or explicit agreements to defer some or all compensation in exchange for potentially lucrative options to buy company stock in the future. There are other, nonmonetary assets that people invest in their companies. They invest time, for example, when they promote products and services in their off-hours as a way to support firms' goals and generate new demand. As I saw in my field research in the New York–based Internet industry, employees spent many of their off-hours marketing their firms at industry networking events, and they talked about the interpersonal connections that they made at such events as a form a social capital for themselves and their companies. Sociologists think of social capital as a kind of asset, and I argue that people can invest their social capital in the companies where they work by tapping their personal connections for information and other resources that, in turn, often provide crucial support for their companies. These ties have been shown to provide critical resources for fast-growing companies and regional economies, but less is understood about how people work to develop and maintain these ties through mechanisms such as after-hours networking. Venture laborers can invest their human capital as well, agreeing to learn and update skills in their own time that could benefit their companies, and this too can be framed as an investment. Skill as a form of investment is one that an employee makes both in herself and in her company or industry. As Hacker wrote, "Skills are not costless to obtain, nor do they come without risk. Skills are an investment, and often what economists call a 'specific investment'—an investment that is tied to a particular line of work, industry, or technology. And the more specific the investment, the greater the cost and dislocation if that investment is left 'stranded' by economic change."[44] Richard Cantillon in his 1755 *Essay on the Nature of Commerce in General* explained that the costs of employing an artisan is more than that of a "common labourer" because it is "dear in proportion to the time lost in learning the trade and the cost and risk incurred in becoming proficient."[45] Cantillon noted that costs of products incorporated the risks of their production, in general, and specifically, the risks of becoming proficient in a particular trade. Risk entered the

Anglo-American tradition of economic thought as a cost to be accounted for, and one for which price or insurance should compensate individual economic actors. But skill is a different kind of risk; it is quite costly in terms of the time required to get training, and for many college-educated workers, there is a range of people—from parents to taxpayers—who bear at least some of the cost of that skill. Learning skills that are useful in one industry makes job loss even riskier. As Hacker phrases it, "The educated rise farther, but increasingly they fall farther, too."[46] Training, especially at one's own cost, is a form of investment that in jobs and industries with highly specialized skills requires enacting venture labor, and this was the case during the dot-com boom.

Another form of venture labor involves shifting managerial responsibility onto the employees themselves. Flexible, short-term, project-based workplaces place more responsibility for getting and keeping work on employees themselves. Heightened job insecurity means workers are increasingly exposed to cyclical economic risk, and flexible workplaces have placed increased managerial responsibility on their employees. As one cofounder of a news web site put it, "I don't want someone who's going to ask, 'What's my job?' I need someone who's going to figure out that on their own." Considering the quick turnaround times on the development of computer applications, employees are expected, in the words of one programmer, to "hit the ground running" with continually updated skills, including new programming languages and familiarity with new technologies. This "individualization of the labor process," as Manuel Castells termed it, aims at "decentralizing management, individualizing work, and customizing markets, thereby segmenting work and fragmenting societies."[47] Being in companies with less middle management and administrative support gives them a feeling of more autonomy in their work and, ironically, a greater sense of attachment to the very companies that have eliminated loyalty within the organizational culture.

While companies may benefit from venture labor investments, the increased job insecurity employees experience encourages them to make these investments in the first place. "The culture of the new capitalism,"[48] as Richard Sennett calls it, replaced a rhetoric and expectation of company loyalty with venture labor and other forms of personal responsibility. The shift in corporate culture from company loyalty and responsibility toward employee risk taking gave rise to the entrepreneurial behavior of the dot-com era. With any job there are inherent risks that workers face given the nature of business cycles. However, working as venture labor means being exposed directly to the economic risk that companies face without the same level of protection afforded to capital investors. Investing one's venture labor presents the possibility of direct or indirect future rewards, but the ability to take risks is not equitably distributed across society, so that people who are already vulnerable in the labor market face other, new inequalities. Moreover, they are directly exposed to market forces and often are without the shelter of company pensions, retraining commitments, clear internal promotion opportunities, and companies' implicit commitments to protecting their jobs.

Like venture capital, venture labor provides valuable resources for companies, new and established alike. Venture labor also creates the discursive and rhetorical mechanisms and language for employees to "buy into" the goals of their companies. Long gone is the expressed conflict between the interests of stockholders and the interests of employees. The discourse of risk gives potent rhetorical power to the tight alliance of employees' interests with owners' interests and company directives. Using venture labor helps employees articulate to some degree a personal sense of "ownership" in their employing companies and in the projects they complete for them.

Venture labor, though, entails risks that are often not as equitably distributed, neatly accounted, or directly compensated as the risks that financial investors face. The risks that ordinary employees take, to continue with the investment metaphor, are often not as portable, as easily diversified, or as fungible as financial investments. Financial capital moves with relative ease compared to labor and employee's investments of company-specific skills, social capital, and deferred compensation. Employees' investments in a company or career often bind them more tightly to *particular* companies and industries, a stark contrast to financial investors' attempts to diversify the risks they face across sectors. Consider the years of industry-specific knowledge and experience gathered over the course of a career. The more specific the preparation for a job, the more closely that skill set is tied to the economic performance of a particular company or industry. Some fields, especially creative industries, require years of unpaid training and internships or work "on spec" before payment can begin. The less likely this experience translates outside the company or industry, the less fungible, or transferable, an employee's investment is.

Venture Labor in the Internet Industry

Venture labor is an investment of people's time and work, bearing risks with potential payoffs in innovation, sometimes with an impact on the

larger economy. One lasting impact of the dot-com era was the introduction of venture labor as an important factor in economic growth, and
during the rise of the Internet industry, venture labor investments fueled
stock market growth for the entire country. The dot-com boom will be
remembered as a time when employee entrepreneurial behavior led market
growth.

That is why the Internet industry is a good place to start to study
venture labor, even though it is by no means the only sector that has
entrepreneurial employees or the only place to find venture labor. In the
Internet industry, people were explicitly exchanging work for future possible payoffs and explicitly engaging with, managing, and discussing risk
within their jobs.

These types of entrepreneurial risks *within jobs* were so unusual before
the dot-com boom. By the height of the dot-com boom, risk taking was a
primary theme of U.S. business discourse. Business magazines such as *Business Week* and *Fast Company* viewed risk taking as something to be embraced,
rather than ashamed of, and published articles on why failing in a venture
is actually a good thing (namely, it shows initiative and experience for the
next venture). Daniel Pink wrote a manifesto for the "free agent," an individual entrepreneurial worker who is "working solo, operating from her
home, using the Internet as her platform, living by her wits, rather than
the benevolence of a large institution, and crafting an enterprise that's
simultaneously independent and connected to others."[49] The media covering the Internet industry during the economic expansion championed this
free agent model of work, and indeed the term was first coined in a magazine that covered the Internet industry.[50] The dot-com boom is also a good
place to explore the cultural dimensions of economic risk. These cultural
dimensions were important for attracting people to the industry as well as
for making those risks seem safe.

For example, in *The High-Risk Society,* a book published at the beginning
of the dot-com euphoria, economic journalist Michael Mandel tells the
story of Kenneth Olsen, founder of Digital Equipment Corporation. Olsen
did not believe that the personal computer was going to be a significant
technological development. He was wrong, of course, but the moral of this
story, Mandel argues, "is he should have been right: the odds of a technological innovation sticking and becoming a big deal are extremely low and
there is no way of foreseeing which one might make it."[51] The Internet
industry is no different. In 1999 and 2000 people made choices—what
Mandel would call "high-risk, high-reward" choices to work for start-ups.
Mandel argued, and I agree, that these types of choices are becoming more

common in the new economy—where more is at stake for a chance at a larger payoff.

In the new economy, ordinary employees integrated this acceptance of risk into their own narratives about the choices they made, regardless of the likelihood for their risks to pay out and even when it was clear that the risks that they took were bad ones. These narratives were at times almost tautological in nature and often prescriptive: risks have positive payoffs; thus, people do and should take them. Within the Internet industry, economic risks and personal hopes were so often intertwined that taking risks edged into the realm of passion, not clear calculation. In the new economy, a person could talk about risks in the same breath as hopes and dreams. To take risks was to believe in dreams, dreams within reach, dreams attainable through work. This was easy to contemplate when the Internet stock prices were at their dizzying heights and risks within the new economy seemed close to sure things. What struck me in the interviews I did for this book was not so much how people talked of their work as potentially making them rich, but rather how "risky work" was often equated with fulfilling, rewarding, or challenging work. Safety and security in the workplace became part and parcel with boring, dull, and uninteresting work. This is a far remove from the way we talk about environmental risks or health risks, and certainly any language that equates risks and dreams is a far cry from the cool calculation that one associates with financial accounting and the realm of "risk management" within the corporate world.

Risk created a sense of choice, oftentimes false, that pervaded tech workers' narratives about their careers. For people who worked in Internet start-up companies, the risks they took represented, in their own words, their hopes and dreams and "only the upside" in the words of a software engineer. But these attitudes and rhetoric about being one's own boss and having control over one's company did not emerge by chance or in a vacuum. The social context for this frenzy and the rush to boldly take risks occurred in the midst of major structural change from an economy in which 30 percent of the workforce was unionized to the wide acceptance of at-will employment. The attitudes toward risk that we saw during the dot-com boom happened in the context of the shift from a workplace where regular, full-time employment was the norm to a growing percentage of the American workforce in nonstandard jobs, many lacking health insurance, pensions, and training. Risk presented a choice when jobs were shifting from the security of regular employment to flexible hiring to meet demand only when times were good.

This rise of individualization and the increased importance of work cultures also led to a change in the rhetoric of work. New economy career guru Tom Peters described the model of the new economy worker as follows: "Turned on by her work! The work matters! The work is cool! . . . She is the CEO of her life!"[52] This new rhetoric of work pervaded business magazines covering the workplace and influenced how people framed their own jobs. During the dot-com boom, new economy workers were represented as young, energetic workers taking chances, making money, and having fun. In keeping with this spirit of revolutionary changes, Jeff Bezos's motto for Amazon was "Work hard. Have fun. Make history."[53]

In many ways workers in dot-coms were "avatars" for the cultural shifts toward riskier work that was under way throughout the economy—they represented, magnified, and reflected the changes in progress while making risky work seem attractive.[54] The risks of the era were not solely dependent upon many different individual evaluations, but rather, in keeping with the pathbreaking work of Mary Douglas and Aaron Wildavsky, culturally and socially informed.[55] This means that people making choices during this time did so in an environment that shaped the perceptions of risks. As we will see in subsequent chapters, many people working in dot-coms wanted to pick the "right" kind of company, one that might have economic payoff, provide for career-establishing visibility, or allow for stability relative to the rest of the economy.

But risk is not optional. Risk taking across the high-tech sector was perceived as being a requirement of working in the industry, not just a lifestyle option. An organizer with the Washington Alliance of Technology Workers (WashTech) said, "Most of the workers in this industry think of stability as dead-end,"[56] and throughout the boom-time 1990s, taking risks was seen as the only way to get ahead. Highly skilled, highly educated, and mobile workers were able to take on and benefit from these risks, and welcomed them as an opportunity for personal challenge and growth. In addition, those who "opt in" for work in this industry (as one employee described the process to me) offered individual-level explanations for and solutions to the problems of risk. Taking risks in their jobs gave them a feeling that they could manage layoffs and economic downturns on their own, and in turn fueled an ideology of financial success as being the result of personal, not collective, actions.

It is important that we think of risk in this sociological way, in order to understand the shift in cultural values in the economy. Mary Douglas argued that the word *risk* in contemporary Western societies now implies bad risks, having moved from meaning merely *chance* to meaning the more

serious *danger*.[57] Douglas rightly emphasizes the shift in the cultural discourse around risk toward an individualistic, rather than collective, notion
of responsibility for those dangers: "The modern risk concept, parsed now
as danger, is invoked to protect individuals against the encroachments of
others. . . . The dialogue about risk plays the role equivalent to taboo or
sin, but the slope is tilted . . . away from protecting the community and
in favor of protecting the individual."[58] She did not study economic risks
directly, but she argued that an analysis of how people view risks can show
how communities understand the risks they face and the processes by
which blame and accountability are distributed within them. Douglas's
cultural analysis has rightly shown that understanding who bears the brunt
of risks reveals the level of individualism within a society, and although
her argument about the cultural framing of risk does not explicitly address
economic life, her observations lead us to consider the culture of individu-
ality within economic life. What we will see in later chapters is that the
concept of risk during the dot-com boom connoted individually made
choices with the possibility of great rewards, even while rising job insecurity meant more people faced economic uncertainty beyond their control
in the American labor market.

To many working in the Internet industry before the crash, work in the
technology field was not risky, but rather seemed a sure thing. Growth, at
a certain point, was phenomenal, leading many people in the industry to
believe that while they might lose a job with one company, they would
never, as Sophie said, "lose all the jobs" in the industry. The case of the
Internet industry exemplifies the extremes of risk taking by employees,
with the industry's rapid rise and fall along with a particularly wide range
of outcomes from long-term unemployment to stock-option millionaire
over a relatively short amount of time. This is a clear case of the attraction
to and the effects of risk taking at work, and this clarity helps reveal the
social forces that shape those risks.

This attraction to risk gave rise to the entrepreneurial spirit of the
dot-com era, and people rushed to join in the dot-com boom, in part, out
of a growing lack of job security in the labor market. The same social forces
that reduced workers' job security encouraged them to align even more
closely with their companies by seeking profit-sharing plans and identify-
ing with the products and services being produced. These cultural changes
accompanied larger economic trends and made the economic risks that
people faced seem manageable and at times even desirable.

There is now a common misconception that the Internet industry was
filled only with "digital hustlers," as one oral history called the pioneers

of Silicon Alley.[59] But many people working in New York's Silicon Alley wanted to make new media and new technology, not fortunes, especially because it was seen as an area that fostered "content," or media, for the World Wide Web rather than computer hardware or software. In a survey of Silicon Alley firms in 1995, 91 percent of them responded that they were in the business of "content design and development" for the Internet—that is, creating the web sites that are on the World Wide Web.[60] Those who were early Silicon Alley pioneers created interesting, "edgy" Internet sites, often on their own time and with their own money.

According to accounts in the media, the Internet industry was filled with people who thought they might get rich quick. However, the majority of the people I interviewed before, during, and after the dot-com crash had sought employment in the industry because it matched their desire for creative, interesting work or because they though the industry would provide a stable future. Some people found opportunities in this emerging industry that simply were not available to them in other sectors of the economy or used these well-paid jobs to support them as they pursued a career in Manhattan's hypercompetitive arts or media fields. People working in the Internet industry reported feeling freer there than they did in other industries—doors were "wide open," as one respondent who had jumped into the industry from book publishing said. The medium itself was "the freest around," as a former documentary filmmaker said of his online job with a corporate media concern. In the words of another person interviewed, job openings matched "talents not resumes."

What economic risks did they face? I want to discuss three directly: the risk in stock options, the risk of unsteady work, and the risk of industry collapse. More than 40 percent of those employed by New York Internet companies in 1999 got some form of stock options or deferred income as part of their compensation package.[61] While to some extent employees have always been exposed to the vagaries of the market, pay tied directly to their company's stock performance is a relatively recent trend. This trend began with CEO compensation as shareholders tied executive pay to stock prices. During the dot-com boom, this trend trickled down the company ladder in technology companies. What began as a movement toward economic incentives for CEOs to keep share prices high has been used by companies as partial compensation for employees who have little direct effect on share price, and there is evidence that cash-strapped start-ups used stock options in lieu of at least part of workers' salary in order to save money on salaries.

The second risk they faced was that of not having steady work. At its height in early 2000, the New York City new media industry had more than 138,000 jobs, according to an optimistic report—more than the area television, magazine, and book publishing industries combined, and just under the number of workers in brokerage and trading firms in New York. Over a quarter of these jobs were part-time or temporary jobs. Even during the industry's most rapid growth period, part-time and freelance jobs grew faster than full-time employment.[62] These are called "nonstandard" work arrangements compared to permanent, full-time employment. Arne Kalleberg and his coauthors found that these nonstandard work arrangements allow firms to hire at a higher hourly base salary in times of tight labor markets without permanently raising salary levels for the rest of their employees. These arrangements benefit employers by increasing flexibility at the cost to employees of stable, permanent jobs.[63] In these nonstandard work arrangements, employees often must provide for the costs of their own training and absorb the time and cost of looking for new work once their project or contract is over. It is in this way that nonstandard work arrangements place cyclical economic risks more squarely on the people doing the work, rather than on the companies profiting from their labor. These risks come in two forms. Contractors and temporary employees, not employers, absorb the risks associated with the ebbs and flows of market demand. They must also invest in skills training and social capital that may or may not pay off. In the New York Internet industry, people spent on average almost twenty hours each week just "staying alive" in their careers through unpaid training and looking for new work.[64] Even at the height of an economic boom, and during a tight labor market for people in technology industries, highly skilled workers faced the pressures of securing steady work. At after-hours socializing at business networking events and parties, people hunted for future job prospects. People in the Internet industry during the dot-com boom welcomed the risks of flexibility as possibilities for new career opportunities, rather than as a loss in job quality. Gideon Kunda, Steve Barley, and James Evans found that highly skilled contractors often welcome this kind of risk as an opportunity to make a personal profit; this is one of the reasons that all flexible, nonstandard jobs are not "bad jobs," as Arne Kalleberg and his coauthors termed them.[65.]

The third risk that people working in the early days of the Internet faced was the loss of work through the collapse of the entire industry or failure of their companies, a risk that seemed impossible to many. With the dot-com crash, those who worked in the Internet industry saw demand dry up for specialized skills, such as programming and project management. They often acquired these skills on their own time and had to figure out how to retool, adapt, or apply them elsewhere once industry demand for those skills fell. Socializing at after-hours networking events—parties to launch new web sites or promote companies, for example—built the dense networks that are critical to companies in an innovative industry, as AnnaLee Saxenian showed in her book about the success of Silicon Valley, *Regional Advantage*.[66] After the dot-com crash, however, many people within the industry found their investment of hundreds of hours of time in building those social networks useless for helping them get a job.

While there were a few dot-com millionaires, the New York Internet industry in 1999 had a median salary of $42,600, which was less than the median salary in either magazine or book publishing. The boom may not have turned everyone in Silicon Alley into millionaires, but the subsequent bust certainly hurt their chances for getting a job. Razorfish, a Silicon Alley design firm, saw its stock market valuation soar to $4 billion during the Internet stock market boom, making it one of the most valuable publicly traded companies in the country, only to later see its share price drop to under $1 by the end of 2000. Massive layoffs and closures, "shake-outs," and consolidations swept the Internet industry in New York and nationally. Perhaps with hindsight the "dot-com bomb" seems inevitable, and with it, "pink slip parties" instead of launch parties. As an article in the *Financial Times* poetically phrased it, workers went from "dotcom to garçon."[67]

At the time, economic risks were made to seem natural or inevitable. People working in the Internet industry expressed that they saw little downside to working in a risky industry, even if they articulated their work as risky. These economic risks were discussed in terms of manageability, since people tried to manage risks in different ways. Instead of fearing risks, people welcomed risks, and the culture of the dot-com sector helped them do so. Companies advertised for jobs as if being a risky young start-up was a benefit for working in the field.

However in the Internet industry, economic risk was widely welcomed. Rhetoric about the potential payoffs of stock options were used to motivate workers, even when, as Alexander Lyon shows in his study of the organizational communication of a start-up company, executives knew these claims to be overly optimistic.[68] In this model of the new workplace, promoted by magazines like *Fast Company*, taking risks was tantamount to asserting independence from corporate control and from boring work. At the very moment when a massive shift toward individualization emerged—with organizations sloughing off responsibility for skills training, health care, and other benefits, and even financial accounting such as tax withholding—an ideology about freedom made risk taking not only acceptable, but, at least in Silicon Alley, highly desirable.

The organization of risk and uncertainty within New York City's Internet industry is important for studying risk for several reasons. First, the industry was one of the first to embrace technologies of distributed work, and it has been theorized that these technologies of distribution also destabilize some bureaucratic structures. Second, innovative practices within the industry, both in terms of new products and of work practices, may influence work outside the industry. Third, understanding the work cultures and social organization of the *production* of information technologies can point to the problems encountered in the adoption of these technologies and work practices, especially within other workplaces.

Many individuals working in the Internet industry in New York, particularly in the early years of the industry, would not be classified as "entrepreneurial" or risk-loving in a classic definition, and yet they directly faced market uncertainties on several different levels. All of the "early true believers" reported having nothing to lose by working in Silicon Alley. The chances they took were utilized by a volatile industry that needed to spread its risk among many different actors within the economy. Job-hopping, outside consulting and professional work, and risk taking are all part of the strategies high-tech workers used to guard against the uncertainty of their industry. In talking about risk, they often connected risks to their chances of finding jobs in the future, either through their creative reputations, the reputation of their company, or careful management of their ability to be in a stable environment.

These risks and workers' strategies to manage them together form a currency of innovation. Individual entrepreneurs and risk-taking employees, exciting new companies, and an industry developing a new class of products are all images of economic growth and economic change. The examples from interview data presented in this book show a variety of approaches individuals use to frame risk taking in their careers. All of the workers interviewed reported feeling compelled to continually acquire new skills needed for their jobs and responsible for tasks such as after-hours marketing of their companies, "networking," and the pressure of "keeping up" in acquiring new skills.[69] This "individualization of the labor process," as Manuel Castells has termed it, aims at "decentralizing management, individualizing work, and customizing markets thereby segmenting work and fragmenting societies."[70] Other forms of individualized work practices fostered a highly individualized sense of responsibility for one's own job stability and career trajectory within the industry. Flexible staffing practices and nonstandard employment arrangements increase flexibility at the level of the firm by distributing part of the burden of cyclical risk among subcontractors and self-employed freelancers, who face risks and costs that a firm has explicitly externalized. Within firms, emergent forms of the organization of work itself rely not on standard hierarchical structures that support risk but on heterarchical forms characterized by "minimal hierarchy and by organizational heterogeneity"[71]—forms hinging on employees' adaptability and entrepreneurialism. The dot-com crash exposed narratives of self in which people expressly individualized the risk of the market. Nikolas Rose has argued that the rise of "governmentality" means individuals take on increasingly more risk and assume responsibility, economically and psychologically, for mitigating that risk, and he concludes that working with risk is one of the fundamental aspects of neoliberalism. In this case, it includes placing blame on themselves for decisions to enter the dot-com industry and for failing to navigate successfully the risks of the market. Whether luck, caprice, hubris, or bad timing, the individuals I interviewed often blamed themselves, not the market or their industry, when they lost their jobs.

Creative, Cultural, and Media Labor

Another form of risk that people took in Silicon Alley was the risk associated with being *creative* labor. Media and communication scholars including Richard Caves, Mark Deuze, and David Hesmondhalgh have written that the uncertain demand in "creative industries" such as media and entertainment means people working in these fields face increasing entrepreneurial pressures and risks. So-called creative industries workers who produce these sorts of intangible goods and services have long faced an uncertain demand for their work—being to some extent, as an old Hollywood adage puts it, only as good as their last picture or project. Because they rely on audiences to evaluate their work, many people in communications—and media-related jobs have "portfolio-based careers," in which they piece together projects judged for their creative or commercial impact.[73] I believe that understanding how media workers produce symbolic, informational, and aesthetic goods and services can help scholars understand how economic value more generally is communicatively constituted and mediated, as more industries begin to rely on this aspect of value.

One concept from this research that is useful for understanding the production of cultural value is the concept of *individualization*, the term for each individual's increasing responsibility for his or her own welfare. In a particularly clear explanation, Angela McRobbie wrote: "What individualization means sociologically is that people increasingly have to become their own micro-structures, they have to do the work of the structures by themselves, which in turn requires intensive practices of self-monitoring or 'reflexivity.' This process where structures (like the welfare state) seem to disappear and no longer play their expected roles, and where individuals are burdened by what were once social responsibilities, marks a quite profound social transformation."[74] McRobbie studied the careers of cultural entrepreneurs in London whose blurred lines of work crossed from night clubs to their companies and back. She and others have argued convincingly that the creative labor force is one of the most visible places to see these changes. Scott Lash and John Urry argue that "cultural industries" provide the template for the rest of the economy to follow—that creative, cultural work has the power to influence how work in other industries is organized.[75] Work extends into more corners of life including social arenas, home, and off-hours.

Communications scholar Terry Flew has made the persuasive argument that service work in general needs to be understood in terms of "transactions that utilize relational, persuasive and semiotic strategies to link the production and consumption of creative content," and this is particularly true for those who create explicitly "creative" content.[76] Creative and cultural labor is speculative and increasingly precarious. The "cultural turn" across a wide array of businesses means that workers are asked to be continually self-monitoring and self-reflexive and seek out intrinsic rewards and motivations for their work.[77] While workers in Silicon Alley were situated simultaneously in the worlds of technology and media, their position was one of creating a new medium. However, within cultural industries, content creation and content distribution often have very different kinds of employment arrangements. As one of Flew's informants phrased it, "Writers drive VWs, and the publishers drive BMWs."[78] In the heady, early days of Silicon Alley, people expressed that they were changing the ways that cultural distribution would occur—and that these changes held out the promise that that writers, artists, and other content producers working in this field could, to build on Flew's work, afford luxury cars as well.

It was this approach to cultural production that differentiated New York's Silicon Alley from California's Silicon Valley. Silicon Alley's industry association was called the New York New Media Association, and many of the region's boosters made much of the fact that people in the area made a creative—not just technological—contribution to the Internet. It was this spirit of creating independent, interactive content for an emerging medium that led Silicon Alley web video entrepreneur Josh Harris to predict without irony or humility on the news program *60 Minutes* that his start-up company, Psuedo.com, would eventually put broadcast network CBS out of business. Even though Pseudo failed, the early potential of the web presaged the fears that Hollywood, television networks, and music studios now have about the enormous power of new media.

Now it is hard to imagine a day without searching for information online—or getting news online, or watching videos on the web, or dodging ubiquitous banner ads and other online advertising. The pioneers of Silicon Alley created the first online magazines or "webzines," the first advertisements, the first online soap operas, the first social networking sites, the first online "webisodes," and some of the first online news sites. Early Silicon Alley focused on creating online video, building communities of users, and experimenting with news and entertainment online. The Internet as we know it today owes much to these early experiments. They shaped in a fundamental way how we now use the Internet and their companies and ideas are still with us, even if some of the particular ventures failed.

Content mattered to Silicon Alley, and content from news, blogs, and videos is still relevant to the Internet—perhaps now more than ever. Content has become infrastructure in the sense that the term is used in science and technology studies: technologies that seamlessly and invisibly blend into everyday life. When Silicon Alley emerged, fewer than 14 percent of Americans were online; by 2002, 66 percent were; and by 2009, almost 75 percent of American adults accessed the Internet.[79] Today, Facebook and Twitter are the surefire hits that Silicon Alley companies TheSquare and SixDegrees, early social networking sites, weren't. Google bought Doubleclick, a pioneering Silicon Alley firm, for over $3 billion in 2007. Silicon Alley company CDNow may no longer be a thriving web-based business, but surely no one now would doubt its strategy of using the Internet to help distribute music. The lessons of Silicon Alley are important not only for what they can teach us about a critical moment in American economic history but also for what we can learn about the creation of a medium through the experience of those who did the work of creating it. And more important for venture labor, these lesson show how entrepreneurial values shaped the Internet. As Fred Turner has written in *From Counterculture to Cyberculture*,[80] an early generation of influential pioneers—not necessarily technical innovators, but cultural and social innovators—can have enormous influence on how technologies are subsequently developed and used.

Methods

Economic life is a perfect arena for field research because the economy links personal values to global structures. In addition to uncovering individuals' relationships to social structures and their discourses about the economy, qualitative inquiry can link observations at the local level to scholars' understanding of these global structures. Although the rapid pace of change in urban Silicon Alley may not be representative of the experiences of most Americans in the late 1990s, much is learned through a close examination of how these individuals navigated economic change. In many ways they were at the forefront of changes in U.S. employment structures, and their experiences presaged ones that have become more commonplace in the intervening years. This book contributes to a view of economic phenomena that treats participants in a field as being shaped by—but not powerless against—larger structural change.

The challenges for this book are to examine the rise of the Internet using cultural, organizational, and societal levels of analysis and to connect the actions and representations of individuals to larger economic structures. Attention to organizational processes can prevent researchers from being able to situate organizational settings in larger social structures; within a globalized economy, this is a particularly dangerous oversight. While a close focus on the work process and dynamics of one particular company can highlight its power structures, examining companies across an industry can help researchers uncover the links between formal organizations and among individuals throughout those organizations. This latter approach is the one I took, choosing to examine work across a new industrial field, rather than delving deeply into the practices at one particular location within that field. This choice offered opportunities to hear from a range of people who were pursuing careers in dot-coms, and without this method, I think it would have been almost impossible to able to uncover a systemic yet nuanced functioning of risk across the industry. However, what is lost in this choice is the richness and depth that comes from following one company, one set of actors, through their everyday practices and engagements in making technology.

My immersion in the field of Silicon Alley began literally in my living room, where my roommate hosted Special Interest Group meetings of the World Wide Web Artists Consortium. Before I studied Silicon Alley, I was living in it, and the parties we hosted became salons for talking about what was possible with the Internet. When Sharon Zukin suggested I write one of my first papers in graduate school on the impact of the new media industry on New York's cultural industries, it was easy to find young people working in Silicon Alley willing to be interviewed. I began field research in 1997, just when the first wave of Internet content companies were becoming commercially successful. The benefit of having lived the evolution of Silicon Alley firsthand was to gain invaluable insight into how the participants themselves framed the multiple communities and networks as the industry grew.

While I did not work in Silicon Alley, I participated in the social events that were important to the industry and that my respondents said were central to their mobility and visibility within the field. I attended office parties, toured companies, interviewed people at work, and hung out at after-hours events in bars, restaurants, and people's homes. Although some of these were the raucous parties that dot-commers had the reputation for throwing (and I faced the occupational hazard of trying to conduct research in such treacherous terrain), many of the events I attended were more like spirited intellectual salons or soirées, with heated debates about the future of the web. My fieldwork included attending Internet industry–related conferences, touring office spaces marketed toward small Internet companies, observing failed and extant companies within the Internet industry, talking with and interviewing industry and company leaders, meeting with industry leaders and New York civic officials interested in learning about in Silicon Alley, and participating in public showcases of Silicon Alley firms. My deep immersion in the field also included participating in the online modes of community that emerged within the industry, following many of the listservs and online newsletters that connected workers in Silicon Alley to one another.

Over the six-year period from 1996 to 2002, I conducted in-depth, semistructured interviews with fifty-four people who worked at Internet-related companies in all position types, from CEOs and founders to entry-level data entry assistants. I also interviewed ten people in one company just before the company's successful IPO. I interviewed people who helped found the key organizations that supported work, networking, and industrial development in Silicon Alley, as well as people trying to get a job in Silicon Alley from other high-tech regions. I interviewed technologists, "creatives," and managers; people with years of technology experience and those for whom their Silicon Alley job was their first. I cast my recruitment net wide, snowballing from my original sample, recruiting from Silicon Alley events (including post-dot-com-bust "Pink Slip" parties), and recruiting interview participants from online Silicon Alley forums. My sample includes millionaires, slackers, temps, self-proclaimed office drones, and everything in between. I asked semistructured, open-ended questions about the nature of work in Silicon Alley, career aspirations, and the culture and norms of the office place and industry.

As part of my study of the social network connections in Silicon Alley, I adopted the method of "network ethnography," using social network analysis to justify the case selection of interview subjects.[81] Having a social network map of the field allows for a richer version of the qualitative researcher's tool, the context chart, which allows for the interrelationships among actors within a field to better understand the context for individual and group behavior.[82] Having mapped the field of Silicon Alley social events enabled me to expand the inferences drawn from my interview data by positioning respondents into one kind of visibility in the field. These data were drawn from a six-year listing of social events in Silicon Alley that listed over eight thousand participants at more than nine hundred events.

This approach also dovetails what we now know about cultural production and technology industries. Work by David Stark suggests that network forms of organization pervade single organizations—that capturing what happens in place within networked organizations "increasingly the unit of action, the unit of innovation, and hence the unit of entrepreneurship is not the legally bounded firm but the networks that span organizational boundaries."[83] The decision I made not to follow a single company, but rather to focus on the industrial arrangement of Silicon Alley helped me see the interrelationships among types of work and types of companies. More important, the ways in which people working in Silicon Alley engaged with venture labor was clear through my comparison of people's narratives from across several different types of companies, in different stages of funding, with varying degrees of financial success.

I also followed and carefully read industry trade publications—both in print and online—as well as researched the corpus of general interest news about the industry. I did not do formal content analysis, but this extensive reading showed me which frames Silicon Alley used to talk about itself, how Silicon Alley was framed in the mainstream news media, and the ways in which work and risk intersected with these two sets of discourses about the industry.

For understanding the historical context, I relied on both print and electronic newspapers and trade publications as well as cultural artifacts of the Internet era such as advertisements, brochures, promotional material, and ephemeral matter and was involved in several efforts of the historical preservation of dot-com ephemera. I also rely on newly constructed archives of digital materials including email trade newsletters, archived dot-com web sites, and job advertisements from the dot-com boom to understand the historical context of the emergence of the Internet industry.

Overview of the Book

The chapters that follow are based on research before, during, and after the dot-com boom in New York City in the late 1990s through the early 2000s. I examine the institutionalization of economic and financial risk and uncertainty within this innovative industry using field research, historical and archival methods, analysis of in-depth interview data, observation of industry events, and social network analysis.

Chapter 2, "The Origins and Rise of Venture Labor," situates the rise of Silicon Alley in the historical context of American postindustrial economic changes. New York's Internet industry, or Silicon Alley as it was commonly called, emerged as an alliance among entrepreneurial work, creativity, and technology and focused primarily on providing services to the media and communications industries in New York. This chapter describes both the historical shift from industrial to postindustrial to so-called new economy paradigms of production in the United States and the ways in which new technology industries were framed as the solutions to the problems of postindustrialism. By the early 1990s, many of the institutional and organizational arrangements for handling economic risk and uncertainty were being replaced to a large degree by individual-level strategies, and this, I argue, helped lead to the rise of entrepreneurialism on the part of individuals during the dot-com boom.

Chapter 3, "Being Venture Labor," focuses on the strategies that people in Silicon Alley used for managing the economic risks they felt they faced. These approaches were tightly linked to personal values, such as valuing financial success, job stability, or creativity, and, in turn, these approaches influenced how people perceived and evaluated the risks they were taking. These strategies, taken together, show culturally individuated approaches to economic risk, providing a protosubstitute for the social institutions that were failing to protect workers from risks. This chapter uses the interview data to build a typology of venture labor strategies that is rooted in this valuation process and shows how people talked about the risks they faced and what communicative frameworks they used to manage these risks. These cultural frameworks encouraged an orientation toward venture labor, even from people with vastly different—and often competing—ways of evaluating the worth of their jobs, which in turn supported the powerful ideology that economic risks people face are the result of their own personal choices. Rather than ask whether these workers were attracted to the riskiness of the industry, my research looks at the ways in which *they themselves articulated the uncertainty of their choices*. In this way, my research develops a more nuanced approach to understanding the existing theories of risk.

Chapter 4, "Why Networks Failed," examines one of the resources that people used to support themselves in risky work. Respondents I interviewed considered their social networks to be their unemployment insurance, hedges for risky ventures and buffers for difficult times. They thought of their friends and acquaintances in Silicon Alley as the fastest and most reliable sources for information about jobs, rapidly changing technologies, trends, and prospects for clients or projects. For those working in Silicon Alley, social networks were considered critical in maintaining their skills, knowledge, and employability. While people relied on them, social networks were not strong or flexible enough to provide support during the industry's recession. These social capital investments show that venture labor can be both a private investment in an employee's career as well as an investment in the company, one that blurs the line between the enterprise and the self.

Social networks provided an important resource for individuals in Silicon Alley for acquiring information and skills. These networks in turn provided valuable resources for the organizations in which employees worked. Employees' social networks also helped foster a sense of community. This sense of community is crucial for building contemporary innovative regional industries. The idea of the community also provided support and, in the downturn, *illusions* of support. This perception of support helped naturalize risk, making the next job seem more like a sure thing and less dependent on market conditions than on community connections.

The stock market crash is covered in chapter 5, "The Crash of Venture Labor," which uses the interview data after the dot-com bust along with historical materials to show some of the problems facing venture labor. This chapter shows how individual venture labor investments were put at risk by the stock market crash and how the crash heightened the conflict between different mechanisms of evaluating worth within the industry and the clash of values such as security, creativity, and investment. In terms of theoretical contribution, this chapter develops theories from economic sociology and the rhetoric of economics to suggest the mechanism for the calcification of individual beliefs and perceptions about the market into economic structures, providing a key conceptual link between social constructionists of the market and economic realists.

The concluding chapter draws lessons from the experience of the first wave of the "new economy" for thinking about media production. In this chapter, I also suggest ways to apply the concept of venture labor to work outside of the Internet industry. The public policy implications of venture labor are enormous. Understanding and appreciating the implications of venture labor is especially important in a public policy environment in which the social safety net has been eroding, limiting employees' ability to take on the risks required for innovative industries. Encouraging and supporting venture labor is paramount for continued economic growth and innovation, and, even more important, for creating sustainable work environments that support workers.

Why Risk Matters

Political, economic, and cultural shifts help explain the entrepreneurialism of the dot-com boom. This book follows the responses of workers in the dot-com industry to those shifts. Shouldering risks has become pervasive throughout the U.S. economy, and I argue that the motivations of people taking these risks are deeply individualized. People experience risk personally, framing their risks in culturally informed but individuated ways. Even though economic choices are shaped by shared social influences such as economic trends, this individualization of economic risk does not bode well for organizing collective, social responses to support work in innovative industries. What this book shows is that entrepreneurial behavior is no longer limited to company founders and financiers and that job losses are experienced very personally, even when economic factors may be to blame. The experience of people who worked in the Internet industry illustrates that risk is socially organized—not a natural or inevitable consequence of economic capriciousness—and understanding the benefits and the costs of venture labor helps situate these individual experiences in a social, political, and cultural context and will help people design better workplaces and make better choices in the future.

Risk gives the appearance of choice, power, and individual agency. As such, risk provides a powerful justification for the lack of security in jobs in the new economy. If anything, capitalism's social innovation during the dot-com boom was to make images of risk and the lack of job security a *good* thing. The strong lure of the rewards to risk during the dot-com era created a volatile situation in which risk taking seems to be a way to have control over the economy. By taking risks, people feel as though they have some control over outcomes in a seemingly increasingly capricious labor market. But this embrace of individual risk taking may hinder the ability to collectively demand and create good stable jobs and workplaces for everyone.

While workers in other sectors of the U.S. labor force may not yet embrace the high degree of entrepreneurial behavior as dot-commers did, they will come to accept uncertainty within the economy if high-tech workers' expectations are any guide. Vicki Smith argues that this is already starting to occur as "uncertainty and unpredictability have diffused into a broad range of postindustrial workplaces, service and production alike."[84] The widespread acceptance of economic uncertainty—and the lure of risk for workers—pose a challenge for the labor movement and progressives to counter. While some companies and workers at the top of the pyramid may indeed be able to convert uncertainty into opportunities for wealth and advancement, the increasing numbers of workers in low-end service jobs and temporary positions without the security of benefits or continued employment will not be as lucky. As income inequality increases, employees' entrepreneurial behavior will continue to be a contributing factor in maintaining this inequality.

For example, almost a third of the American workforce now works outside the standard, full-time employment arrangement, and that includes both low-paid, part-time workers and high-skilled independent contractors. As a group, nonstandard workers are less likely to have health insurance or a retirement plan and more likely to be poor. What these workers also have in common, despite their differences in skill and pay, is that they bear more fully the brunt of economic insecurity faced by companies. Company flexibility is gained at the expense of employment security for workers in nonstandard arrangements and in volatile industries and sectors like technology.

The lure of risk becomes a powerful mechanism pushing people to think of themselves—not their companies, not their industries, and not their economies—as solely responsible for their employment and their economic well-being. This attitude shifts social uncertainty and insecurity to individual calculable risk—risk with potential for enormous payout according to the myths of the new economy. What follows is an in-depth study of the lure of risk in booming industries, which like a siren's song traps people on islands of uncertainty and renders their social safety net even weaker. This context of the economy at large helped shape an environment of flexible and insecure work within the Internet industry.

2 The Origins and Rise of Venture Labor

The year 1994 was a turning point in the history of the U.S. economy. It marked the launch of a "new economy"—and a symbolic end to the old economy value of corporate loyalty to employees.

That year, Netscape Communications Corporation launched Mosaic Netscape, the first commercially available graphical browser for the Internet. It was based on a program, Mosaic, that Marc Andreessen had written while an undergraduate work-study student at the National Center for Supercomputing Applications at the University of Illinois. Before Netscape, most people surfed a bare web of text and lists and asterisks and lines. Writing in the industry publication *PC Week* in 1994, Steve Hamm likened the pre-Netscape Internet to a "vast, dusty library with no card catalog and no kindly spinster librarian to answer your stupid questions" in which users would get lost unless they "know Unix and have the patience of a filing clerk."[1] Netscape changed that, ushering in a view of the World Wide Web similar to what we know it as today, rich with pictures, colors, and sounds.

But that wasn't all that the rise of Mosaic brought about. The company that made it was the antithesis of successful technology companies at the time. Started by a graduate school dropout and a serial entrepreneur, Netscape was young, bold, technically smart, and agile. It was a small start-up company that pursued an entrepreneurial opportunity to create an innovative technology. With $3 million in initial funding from Jim Clark (who had founded the company Silicon Graphics Inc. or SGI, which reshaped how computer imaging was done for Hollywood), Netscape started with just seven employees. By the end of 1995 the company had one of the most successful initial public offerings (or IPOs) of stock until that point—within one day Netscape became a company worth more than $1.07 billion, and the company's market value doubled on its first day of trading. It was also the first stock offering of an Internet-related stock, which technically made Netscape one of the first "dot-coms." Netscape's stunning financial success launched a thousand other Internet start-up dreams.

Compare this to another pivotal event that happened in 1994. About forty miles north of New York City, IBM stood as a paragon of high-tech growth, corporate citizenship, and job stability. One of the world's largest corporations, IBM had revenues of over $64 billion and at one point had commanded fully a third of the worldwide computer market. The year before it had been featured in *The 100 Best Companies to Work for in America*, and it was easy to see why. IBM offered one of the best benefit packages in the country including a pension that covered 85 percent of every employee's final pay and one of the country's first work-life balance programs that included a guaranteed, benefits-paid, three-year leave of absence. As one of the authors of the 1993 version of *The 100 Best Companies to Work for in America* explained about IBM, "People are respected there, and [when] you come there you expect to work there forever, you're not going to be terminated abruptly."[2] A thirty-year employee of IBM described it as a place where people were "very absolutely top-of-the-heap."[3] People spent their entire careers at IBM, and they understood the company was loyal to them and who were fiercely loyal in return. The *San Jose Mercury News* called IBM's approach a "jobs-for-life policy."[4] In 1994, however, that policy ended: IBM officially laid off employees for the first time in its history.

By the time layoffs and early retirement programs ended, IBM had almost halved its workforce from 405,500 people to 225,000.[5] One-fifth of those laid-off employees came back as consultants, a practice that became increasingly common in the 1990s.[6] Making IBM "leaner" matched the image of the company that Louis Gerstner Jr., IBM's chief executive, was trying to project. In August 1994 he said to employees, "We operate as an entrepreneurial organization with a minimum of bureaucracy and a never-ending focus on productivity."[7] Under Gerstner's leadership, the image of IBM went from that of a large stable company with long-term employment to an "entrepreneurial" company. Netscape and IBM represented two different models of work in the technology sector. Not only did the two companies bitterly compete over how their customers would access the web, but at stake was which of their models for the workplace would prevail in the new economy. In many ways, 1994 marked the beginning of the new economy with both the Netscape initial public offering (IPO) and the IBM layoffs, and both show the institutional arrangements, technological developments, and rhetorical shifts that made possible the dot-com boom and created the environment for New York's Silicon Alley. Both companies represent a spectrum of how technology in the 1990s was framed—at once sure, staid, and stable and young, rule-breaking, and edgy.

These events also reflect changing attitudes towards entrepreneurialism in the workplace and in perceptions of job security. In response to IBM's announcement of its layoffs, economic journalist Robert J. Samuelson wrote in *Newsweek* an obituary of sorts for the "good corporation." While Samuelson was careful to point out that IBM's news did not signal the end of job security and stability, he argued that IBM's downfall had special meaning, because it had once seemed "the best of the best" of corporations in how it treated its employees. The symbolic impact was significant: "What's gone is a sense of confidence, a faith that jobs—or careers—are permanent. The anxiety may exaggerate the reality, but it is keenly felt."[8]

The symbolic impact of corporate layoffs was half of the cultural shift that occurred in the early 1990s. With the explosive rise of Netscape's stock value, employees suddenly became partial and potentially wealthy owners of their company. This cultural shift meant new ways of approaching the economic risks of work and was reflected in the new rhetoric of risk. In a very real way, this cultural shift signaled a move away from job security in one company toward one of employability across several companies. This was a shift away from model of professional, stable employment of the post–World War II era as described by William H. Whyte in his classic study *The Organization Man*. Whyte's critique was of the individual being subjected to organizational control so rigid that it robs "him of the intellectual armor he so badly needs." For Whyte, it was the strong emphasis on organizational culture during the 1950s that hindered individual creativity and autonomy; he did not fear "that a counter-emphasis will stimulate people to an excess of individualism."[9] Similarly, in 1985 conservative business writer Peter Drucker worried that safety and bureaucracy were being favored over innovation and entrepreneurship: "Even high-tech people . . . will not take jobs in new, risky, high-tech ventures. They will prefer the security of a job in the large, established, 'safe' company or in a government agency. . . . In an economy that spurns entrepreneurship and innovation except for that tiny extravaganza, the "glamorous high-tech venture," those people will keep on looking for jobs and career opportunities where society and economy (i.e., their classmates, their parents, and their teachers) encourage them to look: in the large, "safe," established institution."[10]

Layoffs in the 1990s like those at IBM made the "large, established, 'safe,' company" and Whyte's "organization man" historical relics. The rise of the new economy turned Drucker's assessment of the economy upside down, as entrepreneurial ventures became more attractive relative to other types of companies.

Naming the New Economy

At the time, these changes in the economy were seen as so major that they required the language of the "new." The rise of the Internet—along with several other forces such as increased global trade and the shift from manufacturing to services—was seen as an epochal break from the past. During the height of the boom, people assumed the economy had changed forever. The fevered activity of the dot-com era has been called a "mania." It was likened to the Dutch tulip craze of the seventeenth century. The advent of the Internet was compared to electricity, the telegraph, and the automobile. It was called a new gold rush.[11]

The dot-com boom made it seem as if these economic and political changes were positive and permanent. Labor advocates for years had pressed for employee ownership, and worker participation decision making. Jobs in fledging dot-coms held out the promise of these. The dot-com boom marked a moment of historical transition between two very different regimes of work, from an era of corporate loyalty and job stability to an environment in which insecurity and personal responsibility for risk predominates. However, the dot-com boom alone was not the cause for this change. The people who worked in the Internet industry during the dot-com boom were on the front lines of changes within the American economy at large, in a new industry, with a new set of communication technologies, and with a new medium for content emerging alongside significant changes to political, economic, and industrial cultures.

I argue that the entrepreneurial spirit that arose in the Internet industry during the late 1990s is an extension of the historical trends that removed many of the institutional and organizational buffers to economic risks. The decade preceding the dot-com boom had much to do with the context of the boom. Political rhetoric in the 1980s and early 1990s that cast growth in the technology sector as the savior of a weakening economy influenced how people evaluated the economic risks of that during the dot-com era. As such, we can't understand the rise of the Internet industry without understanding the context of the history of changes in the postindustrial American economy.

Social scientists have studied the transition from the post–World War II reliance on industrial and manufacturing sectors to a "postindustrial" or service-based economy. The changes that came to be called the "new economy" in the 1990s are a continuation of that transition. The economic shifts are real and provide the context for understanding the shift in the culture of corporate loyalty. This shift had a political dimension as politicians shaped new rhetoric to explain what was happening. In this chapter, I argue that the dot-com boom of the late 1990s through the year 2000 marked a political, economic, and cultural transition that had already begun by the time the commercialized Internet appeared. These political, economic, and cultural shifts emphasized individual responsibility over collective, corporate, and government security and embraced individual economic opportunity. With these changes, entrepreneurialism became celebrated as a means of fixing an economy that was creaking under the weight of aging manufacturing base, losing ground internationally, and struggling to recalibrate a delicate postwar balance of welfare state and corporate benevolence. Rather than being determined solely by the technological change of the Internet, major economic, political, and cultural changes in the United States paved the way for the dot-com era and for the prevalence of entrepreneurial attitudes necessary for it.

The decline in manufacturing jobs in the United States left a cultural vacuum that the rhetoric of the Internet filled. Changes in the economy, in markets, and in political rhetoric all supported the idea that the growth of dot-coms would continue indefinitely and that these new technologies could salvage an economy. The Internet industry in many ways represented the purest of these technological fantasies—unprecedented economic growth without the messiness of hardware manufacturing, industrial start-up investments, or slowness of the "old economy" ways of working. The prevailing message: the creation of a communication medium—the Internet—could drive economic growth and security, create wealth, and revitalize the economy.

What I want to show is that people's responses to the rise of dot-coms, far from being the result of what Alan Greenspan called the "irrational exuberance" of the stock market, was the response of active social agents to these economic, political, and cultural changes. In this moment of massive social transformation, an industry arose that embraced risky jobs, risky start-ups, and risky new technologies. The ways in which people created new narrative pathways to explain their careers and represent the risks they were taking paved the way for emergent discourses of the new economy and for the ways in which people talked about security and risk in their work lives more generally.

The new economy was comprised of three significant changes and a rhetorical shift in discussions of the economy was discussed. These changes also led to the rise of New York's Silicon Alley. The three changes—new institutional arrangements, new technological innovations, and new media experimentation—all combined to create what was widely called "the new economy." These changes were not solely responsible for a rising technology industry technology industry or an "information revolution." If anything, the technological and institutional changes required for the new economy and for Silicon Alley predated the explosive growth of dot-coms. That is, the economic and political changes that were happening in the 1990s were not caused by the rise of high-tech industries, even though politicians, the media, and others closely linked these two dimensions in their discussions of what the new economy meant.

From the beginning the term *new economy* had different, contested meanings. *Business Week's* editorial stance staunchly supported not only the use of the term the "new economy," but also the concept that something had changed so dramatically within the economy that it warranted redefining the economic epoch. The magazine first used the term in 1994 in a special issue on the information economy in an article by Michael J. Mandel in which he argued, "Far more than most people realize, economic growth is now being driven not by services, but by the computer, software, and telecommunications industries."[12] In *Business Week's* articulation, technological advances in communication would drive productivity growth. *Business Week's* economic editor described in 1996 an emerging "New Economy built on the foundation of global markets and the Information Revolution," arguing that it reflected "fundamental restructuring" of the U.S. economy through increased exports and imports and increased investment in computers and communications technology. He listed the core industries in this new economy as "entertainment, education, computer services, communications, and consulting."[13] Mandel and the *Business Week* editors continued to enthusiastically cheer for the new economy throughout the 1990s and early 2000s. At the end of 2000 after the stock market valuations of Internet-related stocks had dropped precipitously, Mandel wrote, "The New Economy is reality, not hype."[14]

New Institutional Arrangements

In addition to rhetorical shifts, economic, political, and cultural changes undergird the institutional arrangements of the new economy; these changes gave rise to the risk-taking, entrepreneurial culture that supported Silicon Alley and the rest of the technology sector.

The key economic change was in the stability of work and employment from the 1970s on. As Robert Reich wrote while he was the U.S. Secretary of Labor, "In the old economy of large and stable industries, job security was the rule; in the new economy it is the exception."[15] Peter Cappelli called this the "new deal" at work, which creates a "market-driven work-force," created by a set of management practices "that essentially brings the market—both the market for a company's products and the labor market for its employees—directly inside the firm." Cappelli argued that once inside the company, market logic pushes out the values of reciprocity, commitment, and equity, ending the "policies and practices that buffered the relationship with employees."[16] By this Cappelli means that corporate culture moved from commitment to workers to a culture that expressly linked market performance to pay and job security. If the old economy work style was a marriage between employees and their companies, then this new employment relationship is as "a lifetime of divorces and remarriages, a series of close relationships governed by the expectation going in that they need to be made to work and yet will inevitably not last."[17] Labor scholar Katherine Stone has called this "a new psychological contract," in contrast to the workers' unwritten expectations of company loyalty in exchange for their own. And *New York Times* economics writer Louis Uchitelle argued that the mass layoffs that resulted created a culture in which workers were seen—and saw themselves—as "disposable."[18]

Innovation alone cannot be blamed for this shift in workplace culture. Before the rise of Internet firms, technology companies were considered stable places to work. Companies like IBM and Kodak Eastman are examples of American companies that were the technology leaders of their time, while offering phenomenal benefits, wages, and job security compared to the rest of the economy. Kodak Eastman offered its employees company housing, in-house health care, productivity planning to minimize layoffs, profit-sharing, and jobless benefits paid out of its own private fund.[19] In this arrangement, bigger organizations paid more than smaller companies, in part because they shared more profits, more equitably, among their employees. One economist showed that before the 1980s 70 percent of company profits were shared with workers, which echoes findings from other economists that profits were more evenly distributed between shareholders and employee stakeholders before the 1990s.[20]

Although there was much debate over whether investments in computer technologies and communications hardware improved economic productivity, these investments were widely heralded as doing so by business leaders and the press and, in turn, fueled increased spending on technology goods and services.[21] In 1996, business investment in computer and telecommunications hardware increased 24 percent, equaling over $212 billion and accounting for almost one-third of economic growth that year. In comparison, spending on *industrial* machinery in all of the United States that year was less than $130 billion.[22]

By the mid-1990s, jobs in the booming technology sector seemed among the best in the restructured U.S. labor market. Growing inequality in wages, increased permanent job layoffs, deregulation of industries, and the loss of a manufacturing base all contributed to a restructured economy that came to be called the new economy. These economic changes meant that more people felt less secure in their jobs and in their long-term ability to earn middle-class incomes, giving rise to a belief in risk taking as the only viable alternative in the economy—and to venture labor.

Political Rhetoric Supporting the New Economy

The political rhetoric of both Ronald Reagan in the 1980s and Bill Clinton a decade later set a cultural stage for the rise of venture labor. With manufacturing in decline, competing political proposals emerged for how to fix the problems of the economy. For political conservatives, the answer involved allowing free markets to function more effectively through deregulation, or less governmental oversight in price levels, rules, and industry structure. The result, beginning in the 1980s, was the deregulation of transportation, communication, energy, and banking industries. Deregulation unleashed "powerful competitive forces on the markets for products and labor" far beyond the formerly regulated industries.[23] Companies "restructured" by selling less profitable divisions and outsourcing "nonessential" business functions to other companies. The corporate layoff, once a temporary measure, increasingly came to mean a permanent reduction in a company's workforce. The number of layoffs skyrocketed: at least thirty million Americans have been laid off since the 1980s.[24]

Political changes during the Reagan and Clinton eras established the tone for the dot-com era. The political rhetoric of both administrations was important because each set the stage for individual entrepreneurial values in the 1990s and the utopian belief in the power of technology to transform the economy. Reagan's political rhetoric helped shift political discourse away from collective economic and social security toward individual economic freedom and entrepreneurial initiative. Clinton's political rhetoric relied on tropes of entrepreneurial drive and technological innovation to create images of a resuscitated U.S. economy—and in the process created the discourses that encouraged venture labor in the dot-com era.

Reagan's political rhetoric equated freedom with economic enterprise and celebrated the individual entrepreneur. Reaganism followed a long line of conservative political rhetoric that linked conservative policies to prosperity and was central to an ideological movement that political scientist Jacob Hacker has termed the "Personal Responsibility Crusade." This movement "has been gaining steam for years, refining its arguments and strategies for challenging the very notion of a shared risk."[25] Reagan connected political freedom to economic prosperity in a way that no other politician had done before, and these rhetorical moves were central to both his electoral platform and his presidency. His political talk chipped away at values of collective security and responsibility in favor of individual initiative and enterprise. Perhaps nowhere is this more evident than in Reagan's famous words from his first inaugural address: "Government is not the solution to our problem; government is the problem."[26] As scholars of presidential rhetoric Amos Kiewe and Davis W. Houck have argued, Reaganomics is rhetorical, "sermonic and ceremonial, simple in its praise, blame, discussion of problems, and their causes and solutions."[27] As Kiewe and Houck put it, "At the root of all economic ills, Reagan identified the federal government. Accordingly, economic rhetoric reduced complex problems to a simplified drama whereby the virtuous people fought a bad government."[28] If government was the force that Reagan blamed for America's economic ills, individual entrepreneurial effort was the cure. In his second inaugural address, Reagan said, "Freedom and incentives unleash the drive and entrepreneurial genius that are the core of human progress. We have begun to increase the rewards for work, savings, and investment: reduce the increase in the cost and size of government and its interference in people's lives."[29] This freedom—and by implication entrepreneurial drive—would revive the economy and lead to innovation. While Reagan did not use the term "new economy," he did imagine a renewed, revitalized economy emerging from these entrepreneurial changes: "From new freedom will spring new opportunities for growth, a more productive, fulfilled, and united people, and a stronger America—an America that will lead the technological revolution."[30] In his State of the Union Address the same year, Reagan said: "We stand on the threshold of a great ability to produce more, do more, be more. Our economy is not getting older and weaker; it's getting younger and stronger. It doesn't need rest and supervision; it needs new challenge, greater freedom. And that word 'freedom' is the key to the second American revolution that we need to bring about."[31]

Of course, Reagan was not the first president to talk about the economy, but he was part of a shift in which economic policy became more central to political platforms.[32] Nor did Reagan invent the concept of American individualism. Reagan's heroic depictions of individualism and his practice of linking entrepreneurialism and freedom rhetorically set the stage for the new economy to emerge.

President Bill Clinton used economic rhetoric to express optimism about the power of the "information revolution" to create prosperity and revitalize a weakened U.S. economy, even as he called on American workers to align their interests more closely with that of the management of their companies. Years before the first dot-com stock was ever sold, the Democratic Party platform during Bill Clinton's 1992 campaign for president used the term *new economy* to describe a trade-off between labor and business. In this trade-off, workers would accept "added responsibilities" and join in "cooperative efforts to increase productivity, flexibility, and quality." In exchange, business leaders would give them "an increased voice and greater stake in the success of their enterprises." New economy in this usage signaled a need for workers to adapt and be flexible in light of an economy that was being restructured. The party's platform, "A New Covenant with the American People," used new economy in a way that held out a promise to workers' that increased entrepreneurial risks in their jobs would pay off with increased rewards.[33]

One of the best statements of Clinton's approach to what he called "a new American economy" was in a speech he gave as a presidential candidate in 1992 at the University of Pennsylvania's Wharton School of Business. In it he laid out broad outlines for an economic plan that would invest in physical and technological infrastructure, encourage closer cooperation between management and employees, and stress increased worker skills and education. Within the speech Clinton used the phrase "new economy" or "new American economy" four times and the word "technology" or "technologies" eight. In contrast, he mentioned the deficit only once. Although he did not mention the Internet specifically in this speech—in 1992 the Internet consisted only of text and relatively few people were online—he did talk about the policies that need to be in place to create a "national information network." He argued that "in the new economy, infrastructure means information as well as transportation," and he proposed a fiber-optic system that would link homes, classrooms, and businesses.[34] Already, candidate Clinton was drawing distinctions between ways of working in the old versus the new economy: "The old economy of a generation ago rewarded countries whose firms had strong organizational hierarchies and strict work rules. In the new economy, our prosperity will depend instead on the capacities of our workers and our firms to change. As Peter Drucker wrote, . . . 'The factory of tomorrow will be organized around information rather than automation.'"[35]

The speech links concepts of technology and information—the "factory of tomorrow"—with the idea of increased cooperation or alignment with the interests of management. For prosperity in the new economy, workers will need to change and give up the "strict work rules" of the Cold War era, and presumably of highly unionized workplaces. Specifically, Clinton called for "a new spirit of cooperation between labor and management that will forge a new compact for economic growth," which he argued was necessary for success in a global economy. This called for "a whole new organization of work, where workers at the front lines make decisions, not just follow orders, and entire layers of bureaucratic middle management become obsolete." In exchange, Clinton suggested restoring the link between company performance to worker pay "by encouraging companies to provide for employee ownership and profit-sharing, and recognize that we should all go up or down together." He drew on the same images of entrepreneurial energy and individualism that Reagan so effectively used and, like Reagan, Clinton quoted Peter Drucker on entrepreneurship. In Clinton's vision of the new economy, responsibilities and benefits were balanced so that

everyone will have to change, and everyone will get something in return. Workers will gain new prosperity and independence, including health care and training, but unions will have to give up non-productive work rules and rigid job classifications and be open to change. Managers will reap more profits but will have to manage for the long run, and not treat themselves better than their workers are treated. Corporations will reach new heights in productivity and profitability, but CEOs will have to put the long-term interests of their workers, their customers, and their companies first.[36]

This political vision of the new economy was in sharp distinction to the realities of corporate downsizing at the time. Clinton the presidential candidate urged Wharton students, future business leaders, to be responsible in caring for labor. Notice, however, that the overall tone of the speech is not that of protecting job security for the anxious middle class or helping people navigate the economic changes that were under way. Rather, the pro-business attitude that would eventually come to dominate Clinton's rhetoric on the new economy was already in place here. Clinton

articulated the shift toward a political culture of risk clearly and force-
fully—"workers will gain new prosperity and independence" if they accept
the terms that "we should all go up or down together."
    The political rhetoric about the new economy echoes what Vincent
Mosco has called a myth of the "digital sublime," in which postindustrialism, globalization, and the commodification of communication converge
to produce an account that functions for political and economic purposes.[37] The digital sublime is the reason why people believed that these
new economy trade-offs and risks would ultimately pay off and that technology could transform society. Such myths "are not just a distortion of
reality that requires debunking; they are a form of reality,"[38] constructing
the communicative landscape for the creation of economic value.

The Cultural Shifts toward the New Economy

Coupled with these changes in political rhetoric was a shift in workplace
culture and business press discourses of financial and economic success.
Deregulation and the political rhetoric around it has been blamed for creating a "winner-take-all" culture in which small differences in performance
or luck lead to growing gaps in salaries or market share.[39] Or as economist
James Medoff said in 1995, "Companies today are financing their major
and minor superstars with what they aren't paying everyone else. . . . The
name of the game is screw the losers to support the winners."[40] At the same
time, performance-based pay increased for corporate chief executives
whose salary was tied directly to the stock market performance of their
companies. Disney CEO Michael Eisner, for example, set a record in 1997
by exercising $570 million in stock options, making him the country's
highest paid executives. Executive salaries continued to boom through
the 1990s, in part due to stock options and other bonus compensation.[41]
The result was more people at the top tying their bonuses and incentives
to stock market performance. This trend trickled down the ranks in entrepreneurial firms, leading to an increased risk taking on the part of
employees.
    As evidence of this cultural shift, CEO speeches emphasized "entrepre-
neurial" values over company loyalty or collectivism. For example, General
Electric's rhetoric espoused "the only job security is a successful business,"
and AT&T's CEO called for a mission to "encourage entrepreneurship,
individual responsibility, and accountability" at the company.[42] Outsourcing and layoffs, however, increased the gap in wages within and across
companies, and created a situation where stable, well-paying jobs were

replaced with flexible work. The process of outsourcing places jobs outside the organization, often cutting off opportunities for people to move up internal job ladders and placing more risk on individual workers.

Another cultural shift was an emerging rhetoric around financial markets. What people say about markets, to some extent, makes markets.[43] Nigel Thrift has called the new economy a "rhetorical fabrication,"[44] and others have argued that to dismiss the rhetorical flourishes of the new economy is to miss the "symbolic efficiency of discourse, the way it structures our experience of reality."[45] Some media representations matter more than others, of course, in this discursive feedback loop of market perception, financial market representation, and continual monitoring of markets by market makers. The *Financial Times* presented itself as "the newspaper of the new economy,"[46] and *Business Week*, which first used the term "new economy" in 1994, became a strong advocate of the concept within the media. An explosion in the number of financial media outlets including twenty-four-hour television channels, increases in the amount of reporting on finance by the mainstream media and the number of advertisements for financial services companies, and the concentration of information in "news, information, and technology" companies like Reuters and Bloomberg helped spur the rise of these narratives about financial markets and economic growth.[47] As Thrift has argued, these narratives about finance were "an attempt at mass motivation, which, if successful, could result in a new kind of market culture—or a spiritual renewal of an old one."[48] Not only did this discourse constitute the market, it was, as Thrift argues, an attempt to reconstitute a new economy itself.

Regardless of whether or not the changes in the economy warranted the label "new," narratives about the financial market framed how technology was viewed. Karel Williams has argued that the new economy both represented "business as usual, acting out changes and continuities which are part of our future as much as our past and which have as much to do with finance and politics as with technology" and reflected the increased financialization of Western economies.[49] While the stock market was booming, new business models and new financial models attempted to explain the growth and attribute causality to the so-called revolutionary changes in the Internet. Although the Internet did and has changed how information and goods are distributed, the effects of these changes took much longer than new economy proponents initially predicted.

In the business press and popular culture alike, the Internet was framed as transformative within the economy, and the label of a "new" economy suggested that these were fundamental and permanent changes. New economy proponents suggested that new ways of thinking about productivity, value, and work across the entire economy—not just in technology sectors—were needed because of Internet innovations. These narratives were prominent in financial and general news accounts of the financial markets in the dot-com era. These were narratives of the inevitability of profits in the technology sector and of the financial markets as simultaneously able to reflect these value and still be incapable of accurately responding to emerging ways of valuation. As Michel Aglietta and Regís Breton phrase this claim, the "inauguration of the new economy had abolished the business cycle and spurred sustainable growth. In effect, the marriage of information technology and free capital markets was supposed to have created a new set of financial dynamics where there would no longer be any trade-off between profit and risk."[50] In other words, in the new economy, financial risk was framed not as risky but as a sure bet, and financial market discourses exhibited this contradiction. The business press reflected this confidence by predicting a historic and continual rise of the Dow: titles of business books published before the crash proclaimed *Dow 36,000*, *Dow 40,000*, and *Dow 100,000*.[51] This discursive move labeled the economy as fundamentally changed, with risk and changes as forces to be embraced rather than feared.

This spirit of optimism in the cultural discourse around the new economy made working in the fledging Internet industry seem like a sure career bet to young workers. An economic moment found its place in the Internet industry, which represented newness, growth, and the clean break from an industrial past. It was then that the creative experimentation happening around digital media in New York began to cohere into an industry.

The Rise of Silicon Alley

Silicon Alley emerged at this turning point in the United States from old economy to new and was influenced by the same institutional, technological, and media changes under way in the economy at large. Just as economic, technology, and media values shaped how the new economy would be defined, so too Silicon Alley served three different masters—innovation in business, technology, and media. New York's Internet industry was best known for its experimentation in content, creating new uses for the Internet as a medium by a workforce that was focused on media making.

Silicon Alley was the name of both the Internet industry in New York City and the name of the district around Broadway in downtown Manhattan where dot-com firms clustered. Small independent companies and projects flourished in a regional network that was supported by organizations like the New York New Media Association and the World Wide Web Artists Consortium. Several trade publications and industry events helped to foster a sense of community. In October 1997, the *Silicon Alley Reporter* published a map of Silicon Alley, which it defined as "loosely the area from 28th Street to Spring Street along Broadway, and three blocks East and West of Broadway along that stretch[52] (figure 2.1). A leading Silicon Alley company, Doubleclick, proud of its status, took out billboard advertising at the intersection of Broadway and 22nd Street across from the Flatiron Building—thus, in the geographic center of Silicon Alley—that read "Doubleclick welcomes you to Silicon Alley."[53]

Silicon Alley was recognized for Internet content design, in part because of its close connections to Manhattan's advertising and media industries. A clustering of companies geographically can create a regional culture that influences both the local work practices as well as media styles through intensive social networking, information sharing, and regional job networks.[54] In New York, content was "king," according to Jason Chervokas, the coeditor of *AtNewYork*, an influential trade publication covering Silicon Alley.[55] The Silicon Alley style that emerged from the early years of this clustering favored high design for Internet content, incorporating writing and art-intensive content for web sites. As Michael Indergaard wrote, "New York City developed a distinctive vision of the Internet as a 'new media' domain."[56] As such, New York City Internet-related companies became better known for innovations and developments in the design, writing, and popularization of content for the World Wide Web than for inventing new hardware or software. Early Silicon Alley was filled with artistic, creative entrepreneurs experimenting with interesting, novel uses for a new medium. In an editorial marking their first six months of publication, *AtNewYork* coeditors Jason Chervokas and Tom Watson helped define Silicon Alley in terms of content by noting the role of New York's "talented artists, designers, writers, programmers, and visionaries to the emerging world of new media."[57] Artistic, creative visions anchored the Silicon Alley community. Even as the variety of projects in Silicon Alley expanded, creativity and arts experimentation was at the core of how the industry talked about and represented itself. Online arts were "more important to Silicon Alley than venture capital," according to Chervokas, because it was "the creative impulses that made Silicon Alley matter in the first place," and losing sight of this would be "a fatal mistake for the New York new media scene."[58] By the time the first industry-wide survey was conducted,

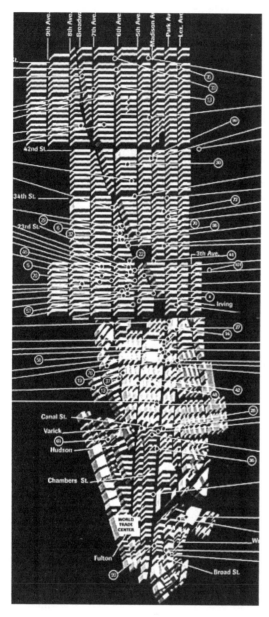

Figure 2.1
Map of Silicon Alley, October 1997. *Source*: *Silicon Alley Reporter*, October 1997

60 percent of Silicon Alley firms were involved in the production and publishing of web content. Advertising and entertainment industries, along with financial services, were the top sectors for clients served by Silicon Alley companies.[59] The people working in Silicon Alley created new ways of experiencing the media, including the first online magazines, the first videos, even the first online advertisements. What Silicon Alley lacked in technical prowess and software engineering, it more than made up for in media savvy. Comparing the creativity of Silicon Alley with the West Coast's technical know-how, Red Burns, a founder of the Interactive Tele-communications Program (ITP) at New York University wrote, "They make machines in Silicon Valley, and machines do not have ideas, any more than a pen can write a novel."[60]

Silicon Alley defined itself as a creative, not necessarily technological or financial, center of the fledging Internet industry. For the early pioneers of Silicon Alley, their work presented the possibility of experimentation in a new medium, rather than technological advancement or financial success. The founders of *AtNewYork* wrote in a column for the *New York Times* that "the Internet revolution is about communication, not technology." [61] While New York corporate media interests were also pursuing interactive technology, the visible faces of Silicon Alley in the media were people in their twenties working in bohemian, artistic conditions who were producing Web sites financed initially with their own money and created on their own time. Calling themselves the "Early True Believers," they created the online magazines music sites and video experiments that defined the New York interactive media scene.[62] A thriving subculture of Silicon Alley–based self-published webzines, experimental online arts and video sites, and early multimedia experiments encouraged a do-it-yourself (DIY) attitude toward online publishing.

These webzines began appearing in 1993 when the World Wide Web was still primarily based on text.[63] Chan Suh, who cofounded Agency.com, created a web site for *Vibe* magazine in 1993, making it one of the Internet's first magazine web sites.[64] Suh's partner in Agency.com, Kyle Shannon, along with his wife Gabrielle, began publishing the pioneering webzine *Urban Desires* from their living room in October 1994.[65] Another independently financed Silicon Alley webzine, *Feed*, was launched in the spring of 1995 out of the apartments of Stefanie Syman and Steven Johnson, who later became a successful business and technology book author. By the end of the year *New York Magazine* called *Feed* "a kind of Gen-Xy online *Harper's*."[66] Marisa Bowe founded *Word* in 1995, which pioneered online video essays. The webzine *gURL*, which Esther Drill, Heather McDonald, and

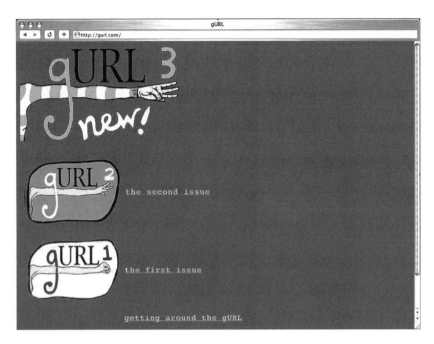

Figure 2.2
gURL.com screenshot, January 1997

Rebecca Odes created in 1996, is a prime example of the creative and alternative spirit that pervaded early Silicon Alley (figure 2.2). Conceived for a class project for New York University's master's program in interactive technology, *gURL* was completely self-financed at first and explicitly challenged the corporate print magazines aimed at teenage girls with proto-feminist coverage of issues such as self-esteem and a positive body image.

Such early experiments shaped how the World Wide Web would develop. These early webzines helped define Silicon Alley style and "first generated national buzz about 'this thing that's happening in New York,'" according to *AtNewYork* coeditor Watson.[67] Once Silicon Alley became known for webzines, investment funding followed. By late 1995, *Word* had been bought by Icon CMT Media, a web design firm that was looking to consolidate several Internet webzines. In 1997, a group of "angel" (or early-stage) investors organized by the New York New Media Association invested $250,000 into *Feed*,[68] reflecting confidence in *Feed*'s magazine model of online content. Other webzines like *Charged* (extreme sports), *Total New York* (urban lifestyle), *New York Online* (urban lifestyle), *SonicNet* (music),

Girls On Film (culture), and *Nerve* (sex) brought visibility to the Internet production occurring in Silicon Alley, and these independently produced webzines helped New York become known as a hub for digital production.

In addition to webzines, early Silicon Alley arts sites like äda 'web, The Blue Dot, and Rhizome expanded the boundaries of digital art production and showcased cutting-edge innovations, bridging the divide between commerce and art. In the words of äda 'web cofounder Benjamin Weil, such sites were "yet another form of research" from which the entire industry benefited.[69] Company-backed artistic web sites provided a creative testing ground for ideas that might have commercial appeal and could, in Weil's words, provide "an opportunity to help artists participate in the definition of this new environment, bringing artists to the core of the cultural revolution this medium could become."[70] Many independently produced experimental projects in Silicon Alley were sponsored by or housed within a larger, for-profit company or agency and blurred the line between creative and commercial. Creative web sites served as a form of research for their parent companies and as a creative outlet or workplace benefit for their employees. For example, Razorfish described its web site The Blue Dot as "a collection of experiences created by the world's most creative artists" that is "ever changing, pulsating, expanding, contracting."

There were similar experiments in building online communities, music sites, and video sites. SonicNet sought to change how people experienced music online. The early online community ECHO first brought together many of the people who went on to found Silicon Alley companies. Pseudo.com, an online channel that showed serialized online dramas and other videos, was a precursor to both YouTube and television networks' streaming video. These collaborations show a close affinity in Silicon Alley between creative and commercial projects and underscore the visibility of arts-driven and media-focused projects in defining a Silicon Alley style.

From Play to Properties

Creative projects like these become a main vehicle for critical recognition, notoriety, and financial success within Silicon Alley. A career pathway emerged in which the independent Web site led to a sort of accidental entrepreneurship, where the founder or founders of web sites profited or benefited from these creations. While few of the pioneers of Silicon Alley set out to become media entrepreneurs, their sites functioned both creatively and instrumentally for building career portfolios, leading to new jobs, creating visibility within the industry, and in some cases generating desirable media "properties" that were bought and sold. Some of these career pathways meant that those who made creative web sites were hired by larger media companies, that their sites were bought or funded, or that they leveraged the visibility of their sites to get hired for other jobs. Personal creative endeavors became invested with an explicit entrepreneurial spirit, and web site founders in Silicon Alley began treating their creative products as if they had potential future payoffs. Without much funding, these experiments were a new kind of sweat equity—with creative investments and risks and efforts to build something with a potential for a future payoff. This fueled the rise of venture labor in Silicon Alley. Although, as we will see in the next chapter, people working on such projects did not necessarily think of their risks in financial terms, they worked within a setting that for many reasons encouraged a DIY entrepreneurial approach to creating a job, a career, and a future within the new medium.

These experiments allowed artists and writers to combine their interests in arts, technology, and creativity within the rhetoric of an expanding new economy that included their vision of a future for the medium. That is, Silicon Alley supported a career narrative that combined the possibilities of creative innovation with financial success, and the community emerging around Silicon Alley reflected and amplified these values of creativity. Many of these early experiments were explicitly targeted toward a specialized audience, as Silicon Alley pioneers pushed the boundaries of the emerging medium. As such, their definition of new media represented a radically different trajectory than those of purely business or financial success. One of these pioneers, Jaime Levy, the creative director of the literary site *Word* and founder of Electronic Hollywood, tapped into this distinction between art and commerce by calling herself a "cyberslacker," which she defined as a twenty-something "who doesn't take too many things too seriously, who sort of plays around with the technology for reasons other than just making wads of cash [for] just pure, unadulterated fun on computers—people that do that, and there's a lot of people."[71] Her attitude reflects a kind of creative career made possible by these early experiments. These careers were not calculated or rational moves. Rather they were attempts at being creative with the possibility of becoming accidental entrepreneurs. Marisa Bowe, the cofounder of *Word*, said of the influential online community Echo: "When we all first started hanging out on Echo, not one of us thought about it in career terms . . . We were all just slackers. To me it seemed like what I was doing [as moderator on Echo] was so useless and it turned out to be a great career move."[72]

However, these early web experiments exemplify what venture labor is. There was low entry into the field and many of the early visible experiments were self-funded. Like Bowe, few of Silicon Alley's pioneers would have described themselves as entrepreneurs in the standard definition of the word. Many of their sites, though, quickly went over a short period from being small and independent to being well-funded with a perceived potential for financial payoff. Websites that were initially created as side projects or experimental projects with little commercial potential in later years proved to be valuable media "properties." Several individually created and managed webzines evolved from entrepreneurial creative projects to mainstream web sites for larger media companies—moving from independent projects to being distributed, as it were, to a larger audience via corporate funding and inclusion on corporate web sites. *Word* was bought by and "serve[d] as a huge billboard for ICon," its parent company, by drawing the attention of potential clients.[73] *gURL* was eventually sold to the teen fashion retailer Delia's, and the founders retained a great deal of artistic and creative control over the site. Delia's used *gURL* as one of the cornerstones for a youth-oriented Internet company called iTurf that had an initial public offering of stock that valued the company at $92.4 million in April 1999.[74] After iTurf failed, the company was folded back into Delia's before the individual web site businesses were sold to other companies. *gURL* was then sold to Primedia, the company that publishes the very magazines that *gURL* was created to challenge, before being sold in 2003 to iVillage. Throughout the site's history, the founders were able to parlay their creative talents into employment with the company that owned *gURL*, whether Delia's, iTurf, Primedia, or iVillage.[75] Like *gURL*, *Breakup Girl*, founded in 1996, became an early independent Internet phenomenon for women, a desirable demographic on the male-dominated World Wide Web. The cable television network Oxygen purchased *Breakup Girl* in 1999; Chris Kalb and Lynn Harris were both hired by Oxygen where they continued work on the site until 2000 (and in Harris's case, also on television programming), when they were laid off and the rights to the use of the Breakup Girl name and characters went into legal limbo. Kalb and Harris obtained the rights to the site and relaunched it on their own in 2003.[76] These examples show how independently created projects coexisted alongside web sites and other Internet ventures that were developed by major media companies.

The First Clash of Values

Even though many popular accounts of the Internet industry leave the impression of a meteoric rise of all companies fueled by bottomless reserves of venture capital or stock offerings, there was an early slump in the history of Silicon Alley. Just as the growth in Internet stocks was beginning, the New York New Media Association (NYNMA) and Coopers & Lybrand survey found a 17 percent failure rate among Internet companies in New York between 1996 and 1997.[77] By March 1998 as several firms announced the Silicon Alley's first initial public offerings of stock, several of the first-wave pioneering content companies and webzines began to close. Five of New York's oldest and most visible content sites—*Word*, äda 'web, *Charged*, *Urban Desires*, and *Total NY*—all closed or suspended publication by the time Silicon Alley online advertising agency DoubleClick held its successful IPO in March 1998. Icon closed *Word*, and the webzine went through a series of false starts before finally ceasing to update the editorial content in 1999. Zapata, a Texas company that had no experience in Internet content, bought the Silicon Alley webzines *Word* and *Charged*, along with thirty other web site businesses, in March 1998 before selling them off in late 1998. March also brought the closure of äda 'web. Digital City, a joint venture between AOL and the Tribune company, acquired äda 'web in January 1997. Unable to fit the site into its advertising model, Digital City shut it down.[78] In fact, content and design became seen as untenable business models for the Internet in light of the rise of e-commerce, and companies scrambled to define their work in other terms. In March 1998, one of the editors of *AtNewYork* called "web shop" the "two dirtiest words in Silicon Alley."[79]

This content slump was difficult for a city in which people had defined their role for the emerging Internet industry as a creative, not technological, capital. Silicon Alley was the place where "getting people off the sidelines [and on to the Internet] is still job one,"[80] and creating content—whether writing, arts, games, or other entertainment was seen as the way to capture existing audiences and attract new ones to the Internet. Just when some Silicon Alley firms were accruing their first sizeable financial gains, the content sector, which at one time had represented Silicon Alley's greatest reputation growth, experienced the first Internet bust. As one observer noted at the beginning of 1997, "It is true that the shakeout that started last spring and continues to this day has taken the elation out being the Internet's 'Content City.' Too much of Silicon Alley looks like a ravaged village the day after the troops have moved on."[81] These closures sounded the "death knell for New York's first generation of Web publishers, the generation of people who built Silicon Alley's reputation but ultimately, not its fortune," according to *AtNewYork*'s Watson.[82] He continued: "What content producers here once offered was ideas—new ways of thinking, experiments in media. That was once the story of Silicon Alley."[83]

Content fueled the rise of Silicon Alley and created the community around which a technology industry was formed. However, the high values that were once placed on creating original content for the web shifted, along with the valuation of the work of making that content. In a passionate defense of the content business as distinct from the rest of the Internet industry, Marisa Bowe argued that webzines are really modeled on magazines, not other technology businesses, and should be considered as such.

One basic misunderstanding that I think most commentators (and investors) have made when thinking about content is to assume that Webzines will follow a very different business trajectory than print magazines do. Because Webzines are hosted and read on computers, they reason, they are part of the computer industry. But that's wrong: Webzines are part of the publishing industry, at least for now, because bandwidth is still low and text is still the basis of most content. And in the publishing business, properties take years before they're considered failures or successes. During those years, investors lose money while the property builds an audience. If all goes well, a loyal, desirable audience is built, and the property becomes a cash cow.[84]

The extent to which she spells out this difference shows how closely linked content web sites were to other, more profitable business models online. Watson made a similar case about the difficulties of Internet content firms being closely associated with more profitable sectors of the Industry: "Competing for eyeballs amid millions of unpaid competitors with a pure content package—no matter how cool, revolutionary, or worthy—does not pay the kind of bucks it takes to satisfy investors, float an IPO, or make the publishers rich."[85] At stake is something that still haunts the Internet as we know it—how content will be paid for while making interesting, creative media. Ultimately, the high values that early Silicon Alley placed on creative experimentation clashed with the demands of the market.

The Role of Venture Labor

Many of these experiments emerged from early creative endeavors but were often later linked to financial business models whether through advertising, sponsorship, or other such means. Silicon Alley's early period of experimentation—and the early days of the Internet industry—brought information to technology, connecting conceptually hardware and software companies with the work of new media, creating the symbolic space for a new industry centered on new media to emerge. The paths of the edgy downtown webzines crossed with those of corporate magazine publishing and advertising already located in Manhattan. Andrew Ross called this process of corporate interests overtaking the Silicon Alley "the industrialization of bohemia." The history of Silicon Alley is more complicated for two reasons. First, while Ross's term adequately describes what happened within some successful content companies, it implies cooptation of independence, rather than a longer and more mutual intertwining of corporate and artistic goals, as was the case in Silicon Alley. Second, the term also implies an earlier, uncorrupted state of independent Internet production. The social process by which Internet production moved from high-cultural elite artistic and cultural production to mass-distributed objects included both the influence of corporate media and that of independent, alternative culture on the norms, values, and aesthetics of Silicon Alley. I would argue that it is better to think about the bohemianization of industry—that new media became a model of the new economy, in which creativity functioned for capitalism. Sweeping technological changes meant that people who understood the Internet's potential as a medium positioned themselves as business revolutionaries. Creative people suddenly found new economic value in their artistic work. And for a cohort of underemployed college graduates, Silicon Alley presented the opportunities of a growth industry. The experimentation, both cultural and economic, in Silicon Alley intertwined with the economy's growing culture of risk so that this spirit of risk taking was flexible enough to adapt to multiple motivations.

Thus, part of Silicon Alley's social innovation was linking technology to creativity. Until the rise of visible online commercial web sites and the growth of the dot-com industry, images of technology work were either that of the staid pocket protectors of IBM engineers or of dangerous, rebellious hackers. A 1995 *New Yorker* magazine cover featured a man with a shorn head, multiple piercings, and a muscle T-shirt begging for change with a sign on his laptop "Will hack for food."[86] While this rebellious image of the dangerous technology expert, fueled by an anarchistic spirit and wearing countercultural attire, continued to be used by those in Silicon Alley, there was for the most part a closer affiliation with chic, high design. Silicon Alley, and the dot-com boom in general, changed these images of work as new ones emerged of young workers who were simultaneously creative and well paid. Several direct references to and citations of the Internet industry in design and fashion magazines linked the culture of technology to high culture—a movement from geek to chic, as it were. Silicon Alley firms also used the cultural capital of high design, architecture, and fashion to associate themselves with successful, expensive creativity. No longer did technology workers wear thick glasses and sensible shoes. People working in Silicon Alley appeared in advertisements for Samsung that were photographed in stark black and white, as if for a fashion advertisement; Rockport shoes used other real-life, Silicon Alley "geeks" to advertise its shoes (figure 2.3).

At the height of the Internet boom, several magazines featured geek-chic combinations of dot-commers and fashion. An April 2000 *Harper's Bazaar* feature on "Fashion Gets Wired" that was photographed at the offices of

Figure 2.3
Shoe advertisement featuring a "geek" from Razorfish. *Source*: *Forbes ASAP*, October 6, 1997, 35

Figure 2.4
Harper's Bazaar geek chic

Silicon Alley company Doubleclick proclaimed, "With today's brave new economy comes clothes to match, as Silicon Valley's high-speed style gives fashion a worldwide wake-up call. From new-media labs to old-money citadels, office attire is evolving faster than an Intel processor, and designers in every style niche are keeping pace with generation Net."[87] The issue featured a fashion spread on "geek chic" that matched high fashion with typical geek accessories like backpacks (figure 2.4) and a spread "Net Work News" that advised "a wild python coat keeps a workstation looking sharp."[88] The following month, *GQ* used the men of Silicon Valley as models, "captured here at their favorite hangouts, in the spring's best sportswear."[89]

Part of this melding of creative and corporate cultures was fueled by the initial public offerings happening in Silicon Alley. The financial interest in Silicon Alley companies meant that companies needed to encourage creative, artistic innovation for the expansion of a new medium, while meeting the increasing demands by clients and funders to conform to corporate values of profitability and revenue growth. Firms did this by trying to create a hybrid of corporate and bohemian values in one organization, as Andrew Ross showed in his study of the "no-collar" work at Silicon Alley advertising company Razorfish. While multiple ways of valuing work presented a capacity for innovation at the emergence of the industry, these multiple values were a liability later, when, in the words of many in Silicon Alley, it was time to get "serious" as opposed to creative. People who were involved with the founding of the Silicon Alley community worried that the loss of a bohemian edge would kill what was interesting, unique, and unusual about both the web and Silicon Alley. One founder of a company bitterly declared "The end of the web as we know it" in a rant during a dot-com party celebrating the industry's financial success. The tensions among those who valued developing content for the Internet, developing lucrative new business models, and developing new technology revealed the fragility of these associations, once built in Silicon Alley. As the editors for *AtNewYork* wrote in "The VCs Don't Get It and It's Our Own Fault," an editorial during the first "shakeout" of Silicon Alley: "We're all still growing up in this industry. The VCs need to understand the value of a content site, and the content creators have to understand the value of a good business model. Let's hope we all find some common ground soon. Otherwise, we'll all end up watching Shockwaves of *Caroline in the City* reruns." This passage points to the struggle between those working for innovative creative production online and those looking for efficient means of distribution of culture to a mass audience. The example is telling: rather than reruns of the network television sitcom *Caroline in the City* (which the authors criticized for being hopelessly removed from the experience of living in New York) being redistributed via the Internet, the author calls for the continued production of original, creative content for the Internet in Silicon Alley, not yielding to the demands of venture capitalists who don't "get it" and would instead make the medium look "vanilla" like the "boring passive media called TV."[90]

The link between artistic production and financial valuation created perceptions that risky endeavors could be evaluated as cool or avant-garde, that independently produced cultural products could suddenly become financially lucrative online "properties," and that start-up firms and individuals could successfully challenge the established business practices and established social networks of corporate media conglomerates. The culture of early avant-garde Silicon Alley content producers interested in pushing the medium to its limits often conflicted with the financial values of those interested in the possibilities of low-cost mass distribution over the Internet. Of course, Silicon Alley is not new in using the images of cultural subversion for mainstream media profit or in incorporating the values of countercultural movements into new technologies, but the youth and financial power made the dot-com hipster a particularly visible urban consumer.[91]

This avant-garde stance was difficult to maintain as the web went main-stream, however. The web site *gURL* was bought by the same teen magazine it was founded to critique. *Martha Stewart Living* published an issue on decorating with technology that featured articles on integrating personal computers into "Martha's" signature traditional style, and in a signal that web publishing was no longer limited to a technological vanguard, the company changed its name to Martha Stewart Omnimedia in an explicit effort to align with online business strategies. As the founder of SonicNet, an alternative music web site, said soon after his company was bought by MTV's parent company, "We have to work very hard to balance the kind of accountability we have to have now with the creative risk-taking that brought us to where we are."[92] The close association of creative valuation with financial valuation emerges partly because of the close affiliation of content creators within Silicon Alley to other types of Internet companies—as well as to corporate advertising, media, and, eventually, financial companies. These processes of association fostered the tight linking of creative and financial risk taking in Silicon Alley, which allowed some workers within Silicon Alley to utilize countercultural, artistic, and creative values to understand their exposure to economic uncertainty.

With these symbolic associations came the risks attendant to different types of business ventures. While some dot-com businesses were modeled on lucrative software companies, management consultancies, and advertising agencies, others were creative endeavors more akin to magazine publishing or arts production. Small independent producers of content for the Internet were not wholly independent of market forces and evaluations, even before the companies they ran were funded by venture capitalists. However, these associations meant that creative risks were equated with financial ones, and both were used in the service of capital accumulation in Silicon Alley. What for a certain group of early Silicon Alley denizens was a situation in which they had little to lose, in which they self-published magazines and self-curated art projects for distribution via the Internet, became a situation in which they risked their companies and often lost their investments in the intellectual property of their Web sites. When Silicon Alley began to experience financial success, it coincided with the destruction of this diversity of company types in the industry and the loss of a sense of community among those who worked there. Just as the first initial public stock offerings occurred in Silicon Alley, the web sites that established Silicon Alley's identity for creative work struggled with definitions of dot-coms that were emerging in the growing demands of the market. In this way, the history of Silicon Alley depicts how diverse business activities became associated as one coherent industry, and how this process eventually exposed the creative content producers to risks they felt they should not have to accept. In the next chapter, we'll look at the ways in which they talked about those risks.

Using a rhetoric of revolutions and social movements, the new Internet industry was both shaped by and in turn supported the cultural, business, and economic changes that were under way in the U.S. economy by the mid- to late 1990s. New York's Silicon Alley was in part the industrial result of a shift in the culture of the economy at large toward more entrepreneurial work and the seemingly endless potential for technology growth. The rise of the Internet industry, and the New York sector in particular, was as if the new economy had found its industry. Venture labor also found its cultural moment. While taking risks at work is not new, it became newly important and found expression in the ways in which people worked in the Internet industry. This chapter has examined the institutional shifts that supported the rise of the Internet industry. Chapter 3 examines people's responses to these changes in how they created narratives about the risk in their jobs that served the purpose of managing that risk.

3 Being Venture Labor: Strategies for Managing Risk

In the last two chapters, I addressed the shift of economic risk from institutions and organizations onto individuals and how the narratives of creativity and renewal supported the rise of an Internet industry. In this chapter, I look at how people are managing those risks through narratives about choice of jobs and through their discursive framing of their careers. People talk about economic risk personally, not as a social phenomenon, and their justifications, rationalizations, and strategies for risk are tied to very personal ways of evaluating the world. Within innovative industries, these ways of talking about risk are how people *become* venture labor, a subjectivity that allows them to access a sense of choice and control, while expressing career passion. Based on what they valued in their careers and jobs, people formulated different strategies for managing the risk of their work in the Internet industry. Some people, for example, valued jobs for their stock-option potential, while others valued work for its creative potential. These evaluations influenced the multiple valuation schemas of worth within the industry. In this chapter, I develop a typology of strategies for *being* venture labor that is rooted in this valuation process. *Values,* such as pride in craft, financial success, and continued employability, structured the risks people perceived and in turn shaped how they responded to those risks. While people within Silicon Alley managed risk differently, this process allowed them to share an awareness of the need to manage risk. Although there may have been tensions among the different evaluation schemas for establishing worth within the industry, collectively they show a process of risk management across the industry that did not require adherence to any one particular framing of risk.

Analyzing the interviews with fifty-four people who worked in New York's Internet industry at some point, I found three coherent strategies for managing risk, which I term creative, financial, and actuarial strategies. These different strategies show the adaptability of the cultural frameworks that enable individuals to understand and rationalize the economic uncertainty they face. The names of these categories are evocative of an approach to risk and reflect the values that shape each of those approaches. These strategies do not map precisely onto the types of people working within the industry—that is, managers do not necessarily use actuarial strategies for dealing with risk in their jobs and lives. I argue that these strategies emerged from the values about work and lifestyle—ways of being within the industry. These styles manifest in three consistent types across the interview data. They could vary within the same company, meaning that these styles or strategies were not a consistent part of organizational culture but reflected instead individual cultural approaches to work within the Internet industry. Taken together, though, these strategies show the process of the individualization of risk through different evaluative cultural frameworks, and these frameworks functioned to socially construct risk.

Andrew Ross argued in *No-Collar* that the "most important influence of the New Economy will be on employees' expectations of work conditions, not on the nature of investment or business opportunities."[1] Dot-commers were "avatars of uncertainty, bruised by the rough passage of market forces, they were also on the front line of economic profiling that was changing the character of work itself."[2] These cultural shifts played out in individual narratives in surprisingly different ways.

Denouncing (and Announcing) Values

I recognized these different values as motivating factors in determining economic risk when I began to hear "denunciations" during the interviews I was conducting. These denunciations of creative values, business values, and financial savvy (or lack thereof) show that there were distinct ways in which people understood their position within Silicon Alley and that different people considered these values authentic or inauthentic ways of being within the industry. Denunciations showed workers staking a claim on authentic lifestyles and debating the boundaries of the industry. Taken together, though, these denunciations point to the ways in which personal evaluations of worth shaped how workers understand and navigate the Internet industry.

These denunciations reflected personal values, or the "worlds," to use Luc Boltanski and Laurent Thévenot's terminology, people use for framing their arguments and choosing schema for economic decision making. In *On Justification: Economies of Worth*, Boltanski and Thévenot argue that different "worlds" provide the justification for values, such as domestic worlds versus market worlds.[3] In an outline of a "critical matrix" in which values from one realm, or world, clash with others, they present the ways in which someone operating from one set of values could denounce the legitimacy of people operating from other sets of values.[4] Denunciations, then, can be examined for insight into the values of people behind the critique. Denunciations are one way we see subjects defining their own values in the negative. Within Silicon Alley, questions of authenticity were hotly debated, and I heard many people within the industry denounce the motivations and values of their colleagues using the language of authenticity: for example, who was "authentically" a part of Silicon Alley or doing "authentic" work. During interviews, respondents would often critique or denounce other kinds of people who worked in Silicon Alley as having an inauthentic or inappropriate stance toward work in the industry. In their worldviews, there were right and wrong ways to *be* in Silicon Alley, modes of subjectivity that fit with how they valued work. The denunciations show how strongly these values were held and how personally intertwined they were into subjects' sense of self. They were ways in which their worlds made sense, ways of viewing and valuing what is valuable about their work, expressed not for my sake but for their own. They also provide evidence of the process of cultural framing that is key for how people manage risk in their economic lives.

These denunciations were expressed in what has been called "style of life," terms that reflect what Max Weber called ways of organizing one's life around a craft to reflect values and status. People invested in their own styles and valuations and downplayed or denounced those of others they saw in conflict with theirs. These styles of life emerged as three relatively stable "group styles"[5] within the occupational communities of Silicon Alley, which in turn structured how people thought about and acted upon the risks they faced. The first type of denunciation was exemplified by David, a graphic artist who founded his own small web design studio. David was contemptuous of people who worked in the Internet industry as what he called a "fashion statement": "My feeling is that new media is not so much an industry as a phenomenon. It's hip to do, you get paid well and in a lot of offices you don't have to work that hard. It's a relatively easy lifestyle that is run by their peers." David identified and then promptly dismissed what he thinks other people in the field are doing. This is a denunciation of those who think that the work is "hip to do" and allows for an "easy lifestyle" with peers, presumably as a young culture industry with a cultural cachet.

Another type of denunciation was used by Jane, a woman who began her career as a writer for new media and worked her way to senior project manager. She emphasized that she entered the field early, and she criticized latecomers who joined the industry for financial gain: "The bullshit stuff that started in 1999, you know, when that took over the web industry in the year 2000. What were all these people talking about? Portals? Metrics? Convergence? Synergy? Meeting people who had MBAs saying, 'Are we all on the same page?'" Jane's criticism was directed at those who used the language of business, and at people who adopted business values within the industry, ostensibly different from those values held by people like Jane who started in the industry first. The terms Jane used, such as "synergy" and "metrics," reflect what she felt were the financial, not creative, values of Silicon Alley. Certainly, these were not the words used to describe enthusiasm for creativity, but potential for new media to expand into different business markets. To Jane, "all these people" using these kinds of frameworks and valuations were wrong.

Internet industry pioneers like David and Jane were not the only ones making denunciations. Mark worked in Silicon Alley for less than six months before he was laid off in 2001. Rather than express anger at the market downturn, he had the following to say about other people in Silicon Alley: "I feel no sympathy whatsoever for people who lost however much money in stock options because that was their own naïveté to be blind to that and to really think that it [stock option wealth] was going to be a salary." To Mark, other people lacked a realist worldview about options and the savvy to see that stock options were truly risky. He prided himself on his realistic expectations and on his ability to figure out what he felt were the "real" risks of working in the industry. And he expressed contempt for those who did not.

The cultural framings that these denunciations point to have ramifications for what the sociologist Lisa Adkins has termed the pervasiveness of the confluence of personhood with property in the new economy.[6] She argues that the social contract model of "control" over personhood in the new economy means that people cannot simply and unproblematically claim to "own and straightforwardly accumulate property in the person, since the relations between property and people are being restructured."[7] This is because, she argues, that new economy ownership and property rights are those of authorship, where it is never entirely clear "exactly when we are quits," and require a reworking of the relationship between property and person.

For people working in Silicon Alley, this meant that they needed to balance professional reputation with cultural values. For example, a senior project manager/writer said that building a reputation in such an environment is hard because companies "pop up and disappear so fast": "I guess I feel like building a career is about building a reputation within a community. The kinds of projects matter, but everybody knows that is a mixed bag—that you don't really have control over what the client wants, and often it's really hard. And there's still not enough money being put into the web for it to really be what it could be, and I think a lot of people are really disappointed. . . . I think in a lot of ways people's goals are really modest—they want to not suck. That's not such a modest goal, but it's a different goal than to be really amazing."

Another way to have control over one's career is to be more entrepreneurial, seeking out new opportunities, technologies, and companies before the market does. One laid-off software engineer explained in 2001 that his retooling was at his choice; he had the choice of which new technologies to pursue, and this gave him a feeling of control over his career despite economic difficulties: "I'm learning wireless stuff. I'm thinking that's the next big thing. There's a trade-off if you're working as a salaried employee. You have to do what *they* want you to do."

The denunciations that people in Silicon Alley used are articulations of their worldviews, ways of framing economic choices that reflect a sense of control and choice—however fragile or untenable. They also show that these workers had self-rationalizations for the choices they made that were at least rooted in a strong articulation of what they thought of as appropriate performance of selfhood through and with good work. The sections that follow describe in detail the strategies that these statements represent—financial, creative, and actuarial.

Jobs as Investments: The Financial Strategy for Risk

Treating one's job as an investment with a potential payoff fits the stereotypical image of dot-com workers. *Silicon Alley Reporter* editor Jason Calcanis called equity "the revolution of our generation,"[8] while iVillage founder Candice Carpenter said "options are the best form of currency."[9] These statements echo what I call the *financial strategy* for managing risk. People using this strategy evaluated their companies for their potential as lucrative investments, accounting for risk in expressly financial terms, and actively assessing the financial potential of their labor as an investment in their companies. People using this strategy explicitly stated the connection between their jobs and the potential for financial success of the companies they worked for and expressed a close affinity to their companies. They believed that working for particular companies presented both risks and the possibility of rewards and that the trick was to pick the company that would lead to a payoff. In this way, they represent the stereotypical Internet industry employee, an aspiring dot-com millionaire. They also embody the extremes of the spectrum of subjectivities of venture labor.

One such worker was Sam, a senior producer at one of New York's top interactive web design agencies, who was starting his third job in sixteen months when I first interviewed him. At twenty-eight, he considered himself successful and largely immune to the industry's volatility. When I asked him what he thought was risky about working in the Internet industry, he said: "I think aspects of it are risky. I think if you're dealing with dot-coms and start-ups now or even a year ago, you'd be stupid to think it wasn't risky. I mean, the technology business carries more risk than other firms. One ways I try to offset it is, I look at it from an investment perspective. If you take the risk, there has to be rewards. Otherwise, why take the risk? So, I think that's one reason I've been able to get on the fast track career, that I've been willing and able to absorb the higher risks and therefore reap the higher rewards." Sam tightly tied the notion of risks and rewards to his career advancement. He considers himself personally adept at absorbing "the higher risks" and thus having a legitimate claim to "reap the higher rewards." He has been "willing and able" to take these risks, ostensibly comparing himself to his contemporaries who were not.

When I asked him how he evaluated potential employers, he replied, as he phrased it, in terms of "tangibles and intangibles":

Well, in terms of the tangible, their business model, how sound is their business? That's one concern I have about this company that just offered me a job. They have cash for eighteen months and after that they either have to bring in revenue or raise more money. It's whether I have confidence in their management. Really, it's what every stock investor would look for in a company. Because that's the view I take. Any company that I want to work for I should be more than happy to invest money and buy their stock. So, confidence in management and I've seen companies that have suffered considerably—[my current company] being one of them—when their management lacks the confidence of not only the investors on Wall Street but also of—and just important—the confidence of their employees.

Sam looked for what he thought "every stock investor would look for in a company," literally comparing his choice of job to a stock investment. For any company that he worked for, he "should be more than happy to invest money and buy their stock." Sam in many ways represents a new kind of worker—someone who actively invests with their labor time, hoping to get a payout for these investments of time within the company. Sam, like many in Silicon Alley, had an individual-level explanation for managing the risks he faced. In his worldview, each individual is responsible for managing risks: "Some people's jobs are more expendable than others, and therefore they're at a higher risk. It's really up to you to manage that risk, to take precautions, and build up savings, what is psychologically and financially [necessary] to handle that [risk] . . . So, in a way, the wind is against everybody in the industry looking for a job right now, because of the market has taken a downturn, but I've managed well, I think."

Sam saw a connection between the larger industry, where "the wind is against everyone . . . looking for a job right now" and his own ability to navigate those changes. To a large degree, Sam's way of taking chances seems to have helped buffer him against the uncertainty facing *particular* firms he worked for. His diversification of his own portfolio of jobs and projects was part of an explicit strategy to stay on the fast track, and staying on what he called the "fast track" helps him "manage well" in periods of market downturn.

For Sam, though, the financial success of the company was not the only way to guarantee his own success. Having what he called a "fast-track career" was another way to manage risk. Fast-track careers, according to Sam, are produced by fast-track companies, which are highly visible companies with prestigious clients and projects. The "real 'wow' factor" on one's résumé, can pay off when obtaining the next job. For Sam, his company's loss of market confidence affected him more through its effects on his reputation and not necessarily through the value or payoff of his own stock options—there was a direct link between the company's perceived financial success or shortcomings (as well as the success of its products or services) and the value of Sam's résumé. The inherent value of his own work notwithstanding, Sam thought that a decline in the market value of his company led to a decline in his own ability to shine within the industry.

Losing that confidence, as Sam thought his employer had, meant that his entrepreneurial investment was simply not worth the risk. Sam's confidence in his company was closely tied to the kind of confidence that stock market investors' might have in a company. The company he was working for and was looking to leave (and did leave shortly after this interview) was one of New York's largest Internet companies. Sam joined the company after it held its IPO, and he was offered stock options as part of a benefits package given to all employees.[10] Sam's calculation about market confidence seemed to reflect a concern for the company's image adding, as he said, a real wow factor to his résumé. However, like other people using the financial strategy as a cultural framework for evaluating risk, Sam judged his company's success explicitly in terms of financial market performance.

Sam's job-hopping stemmed from a strategy to manage uncertainty by getting on the fast track to manage his career. Sam felt "lucky"—his job was good, and compared to more volatile start-up companies, fairly secure. He had "jumped" around, taking more responsibility at different firms to push his job to an even higher level. Sam felt as though having done "a lot of things outside of work, things like publications" and conference presentations made his résumé "look pretty glamorous." He felt he was taking chances by making these moves, which simultaneously put him on the the the path to career success but also dampened his ability to be seen as a good potential employee (and thus to get hired) by more conservative companies. Sam said that he was "unfairly" judged by prospective employers for this mobility, which he considered so critical for ensuring success within the industry: "There's this fight that I want to pick with prospective employers about employee loyalty. I sort of have to answer to why I've made certain moves, certain jumps and changes. And the fight that I want to pick is that that's not really a fair question to ask." That question is not fair, because "no one's job is secure anymore with the exception of Supreme Court justices and tenured professors." He was "lucky and in a good position now" because this work outside the firm helped set him "apart from a lot of other candidates." Ultimately, Sam felt his job-hopping, his outside work, and the risks he took were part of a strategy to guard against the uncertainty of having an expendable job. At the same time, he had little sympathy for less fortunate workers, as they "opt in" for work in this industry, even if they were to different degrees aware of the uncertainties they faced.

John, an early employee at an e-commerce business, expressed how he compared taking risks with an Internet start-up versus working at what he calls a "standard company": "I can work for a few years and have the chance to make millions rather than wait until retirement with a standard company. Yeah, we work long hours but think about all the doctors and lawyers who work the hours we do and only hope to make in the thousands. Sure, I could go to a standard company, put in the hours, move up the ladder get my IRA." This career trajectory, though, for John would not offer the potential for a large payoff—a stock offering that could give him the "chance to make millions." For John, the potential to make millions is contrasted with the benefits ("IRA") and job stability ("move up the ladder") of so-called standard companies.

Alan, an engineer with a sixteen-year-long career in the telecommunications satellite industry, was another person working in Silicon Alley who used a financial strategy to manage risk. Alan came to New York to be one of the first employees in an Internet start-up that developed sophisticated software tools to enable customers to interact directly with web databases. Alan felt he was investing in his company by being the third person hired and having joined the company early. Although it was widely heralded in the trade press as both a technologically advanced and a sound business, the company folded in early 2001. When I interviewed Alan in a café near his Greenwich Village apartment, he had been unemployed for three months and was still looking for work.

Alan chose a start-up out of an explicit desire to be more entrepreneurial. He dismisses "worker bees," steady employees who seemingly lacked initiative or entrepreneurial drive. "Worker bees" have a "standard mentality," and constantly ask "What are we doing this for?" They need direction, motivation, and "management." Alan's own father who retired early from a large oil company was "just a worker bee," and Alan didn't want to end up like that. Stock options helped him feel "more committed" to his company and the long hours, as he explained, were good for his career: "My sister works for a big company, she likes being able to slip between the cracks and not have to worry about the scrutiny . . . do your job and go home at five o'clock. Whereas here the earlier you got in and the later you left. . . . Everybody knew what you were doing. It ruined your life, you know. It's helpful for your career. You can really get a lot done if you work twelve hour days. But what happens in the end? The stock options are worthless and you find that you've just wasted a year of your life." Alan had more at stake than younger workers like Sam. He gave up a career that entailed highly specialized skills and social networks in a different field. With a daughter entering college, he had more expenses to think about than most people twenty years his junior. And he felt it was complicated to get a job as a programmer or a software engineer once one reaches a certain age. Having made his decision based on the investment potential of his labor time at the start-up, he realized that his time at the company was, in his words, "wasted."

There was also a cultural shift to more uncertain work that Alan made in order to have the opportunity for potentially lucrative payoffs. Alan's first thirteen years in satellite engineering were at two of the industry's largest companies, one of which was a defense contractor. The contrast between the stability of large defense and aerospace companies and the instability of start-ups was stark. To Alan, taking a chance on an Internet start-up simply seemed a sound financial strategy, and he explicitly described his decision to work for a start-up as an "investment": "It was nice to work for a company that had a chance at hitting it. That's just like an investment. I mean, that's the same way I invest. I'll put some money in safe investments. But then I'll set aside a certain amount of money to put into stocks that have a one in five chance of making it but if they do they'll make ten times. . . . I've got a couple of those stocks right now. I just like the change." When I asked him what kinds of risks he saw in stock options and what he was hoping to gain, Alan replied: "Just the upside. There's really no downside to stock options. They either expire worthless or they become worth something. It's upside potential. If you don't have it then you're just a, just a worker bee. If you have it, then you're an equity holder in the company and if the company does well then you'll retire early I guess." Even though at the time of the interview Alan was unemployed and had, in his words, wasted years at his former company, he talked about his choice to work there as a good risk.

For Alan, there was certainly an upside for the company in terms of issuing stock options to employees. When asked if equity holding made a difference in his attitude and loyalty to the company, he replied:

Yeah, absolutely. Even looking at the start-up, not everyone was a shareholder. And especially when the company doesn't look like it's going to take off, then you realize your options are not going to be worth anything so it's hard to be motivated. Yeah, absolutely. At [the larger satellite company], we had shares . . . , but it wasn't like they were going to take off or something based on something we did. Whereas, with [the Internet start-up] the things that you did could affect the company and could affect the value of your holdings. So it gives you a sense of power. More of a sense of worth. For working at [the defense contractor] I guess there is a patriotic aspect. But working for them in Silicon Valley with all these millionaires sprouting up, it's a little hard to justify working in a place like that. Even though, you know, if you stay there you can retire at age fifty-eight and play golf the rest of your life. That's what my dad did. He worked for [an oil company] for thirty-five years, and he plays a lot of golf but his golf game still stinks.

Early retirement seemed to be a double-edged sword: had Alan stayed in satellite engineering, at forty-five, he would have soon been joining his father on the golf links. The start-up required long hours that Alan justified to himself and his girlfriend by saying, "I was going to become, you know, rich and . . . it's not like I was going to do it the rest of my life." Like many people using this strategy, short-term sacrifice in terms of their labor was worth the chance for a longer-term financial payoff. Using a financial strategy, Alan hoped that the work and long hours that he invested in a company would be rewarded with a stock market payoff.

After graduating with an engineering degree, Dan had worked for one of the country's largest corporate consulting firms. After a year and a half of focusing on the technology industry, he and his friends founded an Internet software company. Dan raised $2 million in initial funding and had "pitched" his company to "top-notch, first-rate VCs," or venture capitalists. Despite his efforts, the funding and the company lasted only eight months. Our first interview was three months after he had laid off the last of his twenty full-time employees, many of whom were friends. Dan was twenty-four. His charisma and belief in his company was still convincing even though the company had just "gone into hibernation," and the company's demise overshadowed his optimism as he recounted his experience as a founder. He still talked about the company and the decisions that he had made in the present tense, as if talking about something that was ongoing or had happened within the last week.

Dan used the financial strategy for dealing with the risk of being an entrepreneur. Dan invested "heart" in his company and expected the same from his employees. When funding was running low, Dan, who was usually forthright about operations with his staff, hid from them the seriousness of the situation.

I really could have told the team, "Guys, we are in a pretty tight situation. We're walking a tightrope. There could be a big wind and we don't have a safety net." I could warned the team but one, I didn't believe it, that there was a chance that we couldn't get it, and two, you realize how much I stress team and organization and morale. I think that is the lifeblood of a company. I could care less if people work ten hours, eight hours, fifteen hours. It's about how much heart they put into it. The moment you go in and say, "Guys, there's a chance you'll lose your job in a month," they're not going to be that productive in the next four weeks. They can't be, and it's logical. Number one, I don't want them to get discouraged, or two, lose interest in the product.

Those using a financial strategy saw company financial success as determining individual success. Dan and Alan both felt that the investment within start-ups meant that employees could tie their futures to a company's financial success. People using this strategy reported that others, too, were motivated primarily by financial success, and losing faith in the company's possibility of success—even in times where financial failure seemed more much more likely—would "discourage" and "demoralize" ordinary workers. For others, like a recent MBA graduate at a pre-IPO firm, the industry offered the chance to "make it" in a "company that's young and willing to take risks."

For those using the financial strategy, risky investments were those in companies with a chance of great financial reward. What then were the "safe" or value investments for these individuals? For Alan, the former satellite engineer, safety was in having the latest skills: "Even though I'll never use [that particular programming] language again, chances are . . . it's definitely helped me be a better Java programmer. I think a lot of the people felt that way. Even though this job doesn't pay off you're getting skills for that next job. In these articulations, safety was equated with boredom and with limiting success—only through taking big risks could big rewards be earned. People working in this strategy were not simply using rational choice or rational acting frames but rather connecting values about lifestyles to choices about risky jobs.

The concern with the next job or project was central for most people, but those using the financial strategy to manage uncertainty in particular referred to their ability to get the next job as their safety net for being in a risky job. Alan, the former satellite engineer, contended that exposure to multiple projects could help in acquiring skills but could be costly in the long run.

[Building a career on projects] seems like it should work. But right now, this is where I would get most of my material from—do some short-term projects, learn a little bit. Of course, the people who are hiring you don't want you to learn anything. They'd rather you hit the ground running and just do the job . . . and they're gathering intelligence from you and transferring it to their own people who are hopefully up to speed thanks to you. I think that's the big difference between consulting and salary workers. You can do . . . you can expand yourself. You don't have to lie. You don't have to say, "Oh yeah, I can do that." You can say, "No, but I'd like to learn that" and take a couple of weeks and figure it out. But you'd be hard pressed to find somebody who'll hire somebody as a contractor who doesn't already know that.

But for Alan the trade-off with salaried work is that "you got to do what they tell you," which for him meant limiting the kind of research and hands-on training that could prepare him for what he kept referring to in our interview as "the next big thing." When we first met, he was trying to figure out what the next big thing would be. He was learning programming for wireless applications and trying to figure out what opportunities there might be for a new application. Alan calculated the costs and potential benefits of this kind of trade-off and that influenced his choice to join a start-up. Like Sam, Alan "pretty much" thought of his employment decisions the same as investment decisions: Some pay off; some don't. But both Alan and Sam tried to beat the market, so to speak, by thinking in terms of balancing high-risk and value investment strategies. For Alan, value strategies were skills garnered for the next job; similarly, Sam deliberately wanted to diversify his résumé to appear as the most qualified one for his next job application.

Taking chances doesn't seem quite as risky when put into the perspective that skills, contacts, and impressive résumés and projects buffer the uncertainty individuals face within a single company. For buffers to work, however, they need to be transferable.

There are, of course, other tolls of uncertainty. Dan, despite his patent entrepreneurialism and his recounting of the stories of business leaders, described the chances he took in starkly emotional terms. When I asked him what was risky about being an entrepreneur, Dan answered in emotional terms about security: "On a personal level, the risk is not to lose a job. Finding a job is not hard. Most entrepreneurs if they . . . build a company, that person is proactive, has to be optimistic, have talent, creativity. You could put those people into any environment and they would thrive. So, I don't think the risk is from job security standpoint. I think the risk is emotional. You invest a ton of emotion into a venture, into the people, into the product, into the future that when it doesn't happen you risk being emotionally crushed. To me, that is the bigger risk."

For Dan, the entrepreneurialism of new media work was attractive not necessarily because of the promise of large payoffs; in this way, he exhibits a melding of strategies and values to manage risk. As a bona fide entrepreneur, Dan's behavior is the most blatantly entrepreneurial of the people I interviewed. His attachment to the emotional aspects of entrepreneurship, however, provides a stark contrast to the popular image of risk takers. He took a chance, but for what payoff? Clearly money was only one way Dan accounted for success. Value, however, had other dimensions, and with the downturn the differences in these evaluation principles were even greater.

Even though they had fifteen years' difference in their job experience, Dan and Alan were both engineers who had worked in large corporations. Exit strategies from Silicon Alley included going back into industries other than the Internet and creating new ventures. Both Dan and Alan in particular stressed that they could "go back" to their old, more stable positions, and both gave reasons why these jobs were no longer what they wanted. Sam, who was in a stable position at the time of our first interview, soon left to create a new business in the film industry; Dan founded a human resources–related business within four months of our first interview; and Alan was still searching for the next big technology to master that might have an entrepreneurial payoff when I followed up with him. Yet all three acted on a drive to figure out the next big thing and figure out where the investment capital money would be next so that they could know how to invest their own labor.

Those who used a financial strategy, like Sam, Alan and Dan, came to Silicon Alley when it was "hot." They talked the talk of the venture capitalists and mastered the skills they thought they needed to continue making themselves in demand to funders and employers. If they thought of their labor most explicitly as an investment, they wanted to put it where it would have the greatest financial payoff. Diversifying their venture labor investment meant that if one job ended, their earned skills and contacts should lead to another job.

Surprisingly though, the most entrepreneurial employees denominated the value of their venture labor in financial *and* nonmonetary terms. Sam and Alan may have looked at their labor explicitly as an investment into the companies that they were working for, and yet they constructed value around those investments that were not merely about dreams of early retirement or of striking it rich. Even Dan, who spent a year chasing funding at the country's top venture capital firms, offered other justifications for his entrepreneurialism. While Dan said it sounded "cheesy," he said he wanted his work and he wanted his efforts to mean something. Nor did Sam and Alan's job-hopping seem irrational or particularly greedy in a market dominated by job insecurity. While certainly not a new phenomenon, entrepreneurship among employees offers the chance to have more autonomy at work, even if part of the price paid is experienced in terms of "emotions," "credibility," long hours, and insecurity at a particular job. What people using the financial strategy wanted was to have autonomy over what they learn and what they pursue, and they developed strategies and justifications for their actions to insure against the risks inherent in this entrepreneurialism.

People using a financial strategy to manage risk seemingly welcomed the risk of the Internet industry by taking a chance on an untested industry, on small companies with a "shot" at hitting a big payoff, or on rapid job churning in order to land at a higher spot on a career ladder. But even the most entrepreneurial of dot-commers felt pressured by labor market insecurity, their own hopes for autonomy and their own fears about their next job. Those using the financial strategy saw their labor act as explicitly entrepreneurial, while attempting to control the economic uncertainty of the industry.

The single most notable characteristic of the investors is their approach to their careers as investment strategies. They try to "hit it," "strike it rich," or escape being "just a worker bee" through their choice of jobs and projects. They felt that taking a position with a particular company was akin to investing in their career choices. This led to the expression of explicitly financial terms when describing their employment choices.

Nothing to Lose: The Creative Strategy for Risk

People using the *creative strategy* for managing risk articulated that creative projects deserved to take risks, and they expressed the payoff for those risks in terms of project success and visibility, which could lead to future work and career reputation for them. If people using the financial strategy closely identified with the companies they worked for, people using the creative strategy closely identified with the projects they worked on. More than people who use a financial or accounting risk strategy, they connected their values to their projects.

In general, people using a creative strategy to manage risk were often multimedia employees, working across different media and looking for the best possible outlet for their expression. Often they had either creative aspirations or experience in other media such as writing, publishing, and filmmaking.

Lucy was one such person. She also felt that she had "lucked" into new media. She was working as a freelance writer when she and a partner wrote a self-help book, and in 1996 they launched their own independent web site to help promote it. Based on the book, the site spoke to a young, ironic, urban audience and quickly became a cult hit in Silicon Alley. Lucy found herself as the "alter-ego of a cartoon superhero," the web site's virtual star. Lucy and her creative partner were indeed lucky. A well-funded corporate media company bought the rights to the site and hired them both to write and supervise the development of the web site's content. I met her at a "Pink Slip" party, part celebration of being laid off, part job search and networking event. She and her partner had been laid off, and their "baby"— as she referred to, alternately, the web site, its cartoon heroine, and the concept behind it—was in legal limbo. The media corporation owned the rights to the site and the concept. When I interviewed her, Lucy and her partner were fighting to get them back. She emphasized that she was not

a "business" or "tech-side" person, but a "content" person, and she considered her venture into the Internet industry to be an extension of her
interest in writing. For Lucy, her biggest risk was giving up control of a
piece of content that represented her voice:

My situation is a little different than others, because I didn't go into [the Internet
industry] as either a tech person with tech skills to contribute nor did I go into it
as a businessperson. Of course, I have my own business, but not in terms of business
development. I was definitely content. So, for me and for my partner, the risk was
that we were selling our property. The good news was that they wanted us to continue being the people to develop the property which is very unusual and made the
deal very tempting, but considerable. And *considera*ble. But the big risk for us was
what could arise [where we were] selling our property. But that's not unique to a
dot-com. We're selling it. They could buy it and fire us. They could buy it and make
it suck. All that stuff.

The company did buy the site, and in 2001 both Lucy and her partner
were laid off. With the project in limbo, no one was updating the site and
Lucy and her partner feared that the property would simply cease to exist—
without their having access to a claim to relaunch it.

Several strong and visible projects made Lucy a recognizable figure in
both Internet and magazine publishing. She considered a string of projects
to be "all you have" in media industries. Projects were simultaneously "the
currency of this whole business" and what she "had to fall back on" when
she got laid off; they were her identity: "The work I've done, the projects,
the 'street cred' that I had earned as a comedian and a freelance writer. It
was all sitting there waiting for me [after the layoff]. They were not entirely
intact—I had to be like, 'Remember me?' The editors had changed, but no
one could undo the projects that I had done." But projects for those
employing the creative strategy were less explicitly addressed as diversifying job opportunities. For many, projects represented the harried pace of
freelance life and reflected their history of working in other media. Again,
Lucy:

For me the trajectory right now in terms of writing for comedy is recreate what I
had with [the website] . . . That is, to recreate one person, one voice, several media.
I had this advice column. I talked on television. I had a live comedy show. That was
great. Because finally I didn't have eleven jobs—it was all [the website]. So I'm trying
to figure out how to recreate that by talking about on stage what I'm writing. Finding
places who will publish the kind of stuff I want to talk about on stage even if it
means getting a little less money. Supplementing that by sucking it up and doing
projects I don't particularly care about but pay me $4,000. Always calibrating those
choices.

For Lucy, "voice" meant not merely creative freedom, but a brand and a commodity that she could sell. Whatever risks she took when the media corporation bought her site, she was compensated, in part, by the consolidation of her voice across many different media. The company backed her television and writing career around the web site, as she became a personal brand, much like her own mini-Martha Stewart Omnimedia, only Lucy is more urbane, more openly cynical, and definitely funnier. However, her expression of the fear of losing control of that voice demonstrates how fragile reputation can be in creative industries. For Lucy, having one of her projects "suck" would jeopardize her ability to capitalize on the recognition of the quality of that project.

The creativity strategy reinforces the rise of entrepreneurialism. Portfolios of projects are unique, and dot-com designers and writers were often hired for a "look" or a "voice" as much as they were for proven credentials or past experience. Indeed, much was made by many of the writers and designers about establishing *their* look or voice, *their* contacts, or *their* outside projects. Lucy's "outside" project, the web site, was what propelled her into a more corporate and stable setting from freelance writing. Until the site was bought, it had a "Hire Us" page featuring Lucy and her writing partner. Jill, an interactive producer, ran her own web site that helped, as she put it, "bring traffic" to her résumé. And Jane, the senior project manager, said that she felt her own neglect of her personal web site had lost her potential job opportunities.

Many of the early adopters were protective of what they saw as "their" turf, territory that they had carved out for themselves by working on independent or self-published projects on the World Wide Web. Jane's start in the Internet industry came through developing an in-house, online newsletter for a multinational foods company, which led her to work with several different advertising agencies in Silicon Alley. Lucy's independent web site created by her and her creative partner was bought by a media conglomerate. Jill, an interactive producer, got her start in Silicon Alley when her short experimental videos caught the attention of an online agency owner. For those using a creative strategy to manage their careers, independently produced and self-published projects online became business cards, portfolios, and training grounds.

Creative workers in media industries often feel pressure to differentiate themselves from their fellow job seekers. For some people utilizing a creative strategy to manage uncertainty in Silicon Alley, it meant differentiating themselves as "early adopters" of Silicon Alley. That is, they claim to have noticed and worked in the industry before it became a phenomenon of venture capital funding, launch parties, and marketing events. And other people in Silicon Alley were often dismissive of this avant-garde stance. In this manner, those employing the creative strategy often denounced those who entered Silicon Alley later. Jane, a senior project manager and writer, was very careful to introduce herself as an "early adopter" of Silicon Alley and prided herself in being a part of the early phase of new media's growth in New York. More than half the employees I interviewed were so-called early adopters of Silicon Alley, entering the industry before 1997, and they were often dismissive of those who entered later in order to make money. The comment that Jane, the senior project manager, made denouncing business values at the beginning of this chapter is evidence of the tension between those who felt an authentic stance of creativity in Silicon Alley versus those who used the language—and strategies—of the business world.

Many of those using the creative strategy said that they felt they had nothing or little to lose by taking chances in Silicon Alley. Their previous creative and media jobs were far from lucrative, and they told me that their jobs lacked advancement opportunities. In many cases, they held paid jobs in fields that had nothing to do with to the creative careers that they wanted. For writers and artists in Manhattan, Silicon Alley raised their pay rates, suddenly making creative "talent" more in demand than it had been in decades. Dot-coms, even in the early days, paid more and provided more job security than the relatively low-wage service work, freelancing, or entry-level creative employment that many of the workers in this category were doing to support their artistic careers. But as two other interactive producers at two different international media and entertainment (and, as distinguished by my respondents, "old media") companies said, there was another attraction to being involved with a new medium. The Internet industry presented the opportunity to be "a part of making something new" and to help shape the "freest medium around," according to two other people using the creative strategy. Thus, creative strategies often centered on shaping the medium or the product, and those using this strategy attached their valuation of worth to the products and medium they were creating.

Jill, the interactive producer, started her career in experimental film and worked her way up through several Silicon Alley agencies. Her last dot-com job was with one of the largest online advertising agencies and she managed Web site design for several corporate clients. When I interviewed her in 2001 she had made a switch back into old media—working to design content and build community for a radio show. Even then, Jill framed her choice in terms of her creative values saying that "risk is being creative," and emphasizing that she had turned down an IRA and the benefits of a full-time job in order to take what she termed a good creative opportunity.

For others the creative strategy meant pursuing challenging work, even at the cost of job security and money. Kyle Puccia, a recruiter for Silicon Alley firm EarthWeb, described in 1997 how tough it was to lure programming talent away from established corporate firms: "A lot of companies will offer so much money that it's difficult to compete. . . . If you want to be challenged, then work with a company like EarthWeb; if you want to make $100,000, get a boring job on Wall Street."[11]

Workers in Silicon Alley who had skills that were in demand in other industries—like the programmers Puccia was trying to hire—often articulated the creative challenge as one of the driving forces for taking jobs in Silicon Alley. Just as "creatives" such as Jill and Lucy put their writing and artistic visions first, so too did technical and managerial talent who navigated their careers in Silicon Alley using a creative strategy for dealing with risk. Creative strategies were not limited to creative talent.

People using a creative approach to risk did not necessarily ignore financial security and job stability, but they valued job challenges and creative potential more. Jane described this process of evaluation well when she described switching from traditional forms of publishing to new media:

I was more scared about the risk of changing careers, initially. It seemed really risky to me, even though I was really bored in magazines and in publishing. I was so bored I didn't want to do it anymore. I didn't want to do any more books, any more articles, I didn't want to talk to one more person. It was more like a feeling that I wasn't allowed to do something that was going to make me happy. It was much more a personal thing, an emotional thing than something rational like putting food on the table. In terms of the risk, I was always one of those "Do what you love and the rest will follow" sort. . . . I didn't see how the web *couldn't* amount to something, although at the moment, I'm inclined to think that it's the citizen's band of the 1990s.

Jane echoes the sentiments of many using the creative approach to risk: while pursuing interesting work was their primary motivation, they felt that the web would be able to provide adequate financial support and job stability. For Jane, there seemed to be no way the web "*couldn't* amount to something" when she entered the field, although by the time of the interview she likened the Internet to an outmoded niche communication medium, a CB radio.

Creative strategies for managing risk connected people's articulations of the importance of creativity to the choices they made in their careers. Their values of creative freedom and expression allowed them to frame how they made choices or what kinds of companies were "better" than others to work for. Their career strategies emphasized building and protecting their creative reputations and portfolios as a way to ensure continuing work—the next job—within the industry. The next strategy, the actuarial strategy, takes a very different approach to continued employability within the industry.

Don't Count on Anything: The Actuarial Strategy for Risk

If the financial strategy entailed being closely tied to the success of a company and the creative strategy to that of a project, the *actuarial strategy* entailed actively calculating a degree of riskiness for each position, project, and company and then seeking save havens from the volatility by staying keenly focused on career longevity. Those using an actuarial strategy actively sought out stability in their careers and tried to account for the risks they faced. Many demonstrated an attachment to working in the field and wanting to figure out a way to stay in it, despite their perceptions of the volatility they faced. Many reported that they were in the industry because it was what they enjoyed doing, even if it meant figuring out ways in which to keep their positions secure in difficult times. Some enjoyed "sitting around solving puzzles all day" in their jobs or the work of creating web sites. Others reported that they enjoyed the work environments of Silicon Alley. When interviewed, people using this strategy told me about how well they had planned for the stock market crash, or how they had successfully predicted the economic downturn, even if their well-laid plans had not worked.

In that vein, people using the actuarial strategy considered it necessary to manage the risks associated with the industry in order to continue to stay in the industry. Like everyone else in Silicon Alley, they took chances and felt they had an individual responsibility to manage that risk. But their approach to risk was marked by their perception that they had skills that would be transferable outside of Silicon Alley, by their trying not to be too closely tied to financial booms and busts, and by their diligence in researching risks and opportunities within the industry. Like those using creative strategies, they were not working in Silicon Alley for investment purposes per se, but unlike those using a creative strategy they did not define themselves and their work necessarily in opposition to the business values of the industry. While they do not fit the stereotypes of dot-com entrepreneurs, those using this strategy managed to balance the demands of creativity and professionalism more successfully than people using either of the other two strategies.

Those using the actuarial strategy were likely to denounce the "naïveté" of people who did not plan carefully for company failure or the stock market crash. Mark, a twenty-eight-year-old junior producer, who was quoted in the beginning of the chapter, left what he called a "promising career" in book publishing to work in his first new media job. He had been to a prestigious summer program for book publishing and had worked at some of the top trade presses in Boston and New York.

He did his research carefully on Silicon Alley companies before deciding to apply for a position with a teen e-commerce site. The company was seemingly stable and when Mark began his job search it looked, in many ways, like a sure thing: the company was profitable, was growing rapidly, and had an initial public offering scheduled for the spring of 2000. Mark wanted to "go someplace with more opportunities, someplace that could lead somewhere else, somewhere where I'd be making a little bit more money" than book publishing could offer. Compared to the book publishing field, new media was opening doors, and using this metaphor he described the career opportunities in new media: "[In book publishing] I was walking further and further down a narrowing hall, I felt what I was looking for to walk into a room that had a thousand more doors. Whether that was a little more the case in January 2000 than it is now. . . . And I certainly succeeded in ditching the career track, as I'm now sitting here unemployed."

Like other people working on the "content" side of the Internet industry, Mark came into the field with experience in other media, books in his case. Unlike most people using a strictly creative strategy to manage risk, Mark "researched" his choices quite carefully. He also felt he was risking the stability of a full-time, permanent job within a creative field, albeit one in which he felt he had limited opportunities. The main risk that he perceived was to his career:

For me, there was only one risk and that's exactly what happened. That I left a career that was going well in another industry, and now I'm sitting here at two in the afternoon talking to you. . . . To me, what other risks were there? No money was being taken out of my pocket, so there were no financial risks, and there were no emotional risks. I left a career that was going well, and obviously, if I chose to go back into book publishing it might take a little while, but I might be able to do that. I can't [go back easily] for (a) my editor's not there anymore and (b) I'm beyond that level—I'm not going to go back and be an editorial assistant. To that extent, the risk isn't even so huge that I felt no doors were ever really closed. If I wanted to go back, it might take a while. I might not be in as good . . . a position as it was had I not left.

Mark held onto the idea that a job would be waiting for him in publishing if he wanted to return. Similarly, the cost of going back doesn't seem worth it, even at "two in the afternoon" as we sat in his apartment on a workday. Not being tied down was not just about the freedom to take extended trips to Asia, as he was preparing to do when I first interviewed him, but also not wanting to spend a career moving sequentially from "this step, this step, this step." Mark saw that flexibility was being offered in Silicon Alley, even if his stay there was short. Publishing offered no flexibility to move out of a proscribed career track, and Mark felt the pyramid structure of opportunities was bottom-heavy: That editors had "real luxury" in choosing among "dozens" of young, eager college grads for each relatively low-paid position as an assistant, at least when he entered the industry. These were the only ways into publishing and Mark, along with his cohort of would-be book editors, took them with the hope that the career ladder would offer some eventual payoff. For him, the only thing he had to lose by going dot-com was a hard-earned spot on publishing's career ladder, a ladder that was stable, but seemed too long with too many steps compared to the more flexible ways to the top of the Internet industry.

Mark's timing for a career move could not have had been worse. His carefully researched foray into Silicon Alley came just as the stock market downturn began, and the company's planned IPO never happened. As for taking risks, Mark did not invest any of his own cash in the company, nor did he work in the proverbial garage for a young start-up without cash flow, a product, or a solid business plan. In retrospect, Mark was quite pessimistic about the possibilities of stock option wealth:

It wasn't like I was naïve to what the chances were. I mean, I wasn't looking for it to hit. I never—other people—the stock options were meaningless to me. That was Vegas roulette to me. Being realistic, what were the chances of that hitting? At [his company] they were planning on going IPO around the same time I was starting, and even then I wasn't getting too excited about it. Which was subsequently completely canceled. So I wasn't like, "Oh, there goes the down payment on my apartment." I never counted on that anyway.

Question: You were offered stock options as part of—

Yeah, and I took them, but I was offered relatively few at a relatively high price. I think tech companies in general were getting away with murder offering stock options and a lower salary rate and essentially just giving employees worthless paper.

While Mark said he didn't buy into the stock option frenzy, he thought it happened because there were "all these people who were in their twenties and early thirties who were in a position where they were young enough to take a risk to, you know, to go for it." His lack of sympathy could be seen as tied to his perception of himself as non-naïve, as a careful evaluator of his alternatives.

Mark, like many of the other people identified as using an accounting approach to risk, worked in Internet start-ups that operated on a cash-flow basis. These businesses were less likely to be dependent upon large sums of venture capital for product development and instead relied upon business contracts to fund the businesses. Wayne, a programmer, refused to define the company he was a partner in as being in Silicon Alley: "We're not a Silicon Alley company. Those guys downtown, they're in Silicon Alley. We just make boring websites for businesses. We're not going IPO anytime soon. We don't go to those parties." Wayne's company fits solidly within the definitions by both industry groups and the city as what constitutes a Silicon Alley firm, because it provided Internet and intranet design and programming for financial services companies. But being outside the venture capital cycle and "those parties" among companies explicitly seeking reputation as dot-coms, Wayne didn't see that his company fit the image of what Silicon Alley was supposed to be.

People using the actuarial strategy thought, like Mark, that it was naïve to expect that stock-option riches would pay off. In a similar vein, many people felt that they needed to be coolheaded realists about their companies' prospects for stability and their own responsibility for finding stability in their careers. Wayne commented on how strange it was that the programmers who worked with him continued to stay with the company, even though the future of the firm was dim: "My company survives despite having almost no money. I have no idea how. I mean, I know how: just enough money comes in that we can make payroll by cutting the salaries of the executives (meaning me, sometimes). But we told everyone that we can no longer guarantee payroll more than two weeks in advance. And no one quit. The market is getting tighter for programmers, I guess." Wayne justified the fact that his fellow programmers continued to work for the company by assuming that they must not have been able to get jobs elsewhere. The reason programmers stayed on, he argued, is not because they were committed to making the company work, as someone who used a financial strategy might have argued. Nor did he give the reason that people were trying to finish a project before the company folded, as someone using a creative strategy might have said. Wayne believed that if the company looked unstable, then rationally acting others should be jumping ship.

Similarly, Tammy and David took pride in running a web design company "as tight as our design." When I first interviewed them in 1996, they had left their jobs with a premier graphics design and web development firm to found their own small company. Although entrepreneurs, their approach to their business was cautious. Both were trained as artists, and they saw their founding of the company as more a way to determine their own working conditions than to get rich. They, too, were dismissive of those who thought that they could get rich in Silicon Alley. Along with the other people who used an actuarial strategy to manage risk, they belonged to a group of workers for whom the Internet industry represented an expansion of their opportunities, but not necessarily the chance at stock market payoff. Tammy and David did not seek venture capital for their company, and they ran their web design company more like a small business services firm than a potentially lucrative stock offering. They are still in business, and they are not alone. Many argue, like Tammy and David do, that in order to stay in the industry, you have to "make your own work."

Those using this strategy have a clear sense of responsibility for their own economic stability. Of those laid off, they would report that it was their "fault"—that these conditions could have been prevented with more work, more research, better planning, or more drive on their parts, as Evan does here: "It's my fault in a way. I wasn't fast, focused or driven enough. I've always wanted to have a life and in that way it's my fault. I had just broken up with my girlfriend, needed to find a new apartment, and I didn't have the energy to get involved in the politics." Evan had tried to figure out where to, in his words, "ride out the shakeout" that he predicted would happen in Silicon Alley. After working on several projects within Internet-only companies, he took a job with a television network, handling its online strategy. Had he been, in his words, "fast, focused or driven enough," he might have been able to avoid being laid off.

Unlike those using the financial strategy, people using the actuarial strategy had professional aspirations that didn't necessarily involve companies offering stock options. Unlike those using a creative strategy, they saw their responsibility as keeping their options open in other fields and in other sectors of the Internet industry. Joan, a young strategic planner, said that she would go back to law school if new media didn't work out for her. Wayne had been working in the publishing arm of a financial services company before joining a Silicon Alley firm and said that he could go back into that field if he lost his job. Aaron was an established writer for financial magazines, before joining an online publication, and tried to keep his options open in print journalism. Unlike those using a creative strategy, people approaching risk with the actuarial strategy felt they had more options available to them in fields that appealed to them. And unlike people using the financial strategy, they attempted to find safe haven in companies that were less volatile or in positions with more stable working conditions. Evan sought out work with a television network as a way to find a safe job working with digital media. As he said, "I thought in a big corporation I would be buffered from the bubble, but there were other things involved like the new technology that our group introduced proved to be very popular, got all the kudos, but when resources were switched to another team, I knew it would be hard to keep up the pace of innovation that had impressed everyone." Finding these safe havens was one way people tried to manage the risk of the industry.

The actuarial strategy for dealing with risk is characterized by a sense of being able to solve uncertainty as a problem. It is a hybrid, of sorts, of the other two strategies, and people using this strategy felt both an attachment to the work in Silicon Alley and to financial security even in light of the uncertainty of positions in their chosen profession. People using the actuarial strategy attempted to control the financial risks they saw by managing, planning, and balancing their choices to find ways to deal with those risks themselves. By invoking a realist worldview in their choices, they were not necessarily being strictly economic rational actors. Rather, they positioned their employability as something that they had significant control over—that they had the power to make choices that were smarter, savvier, and "better" than people using other ways of being and working in Silicon Alley. By doing the right things, they thought they could outwit and outsmart markets—finding the expanding sectors of the industry and making for themselves a stable career pathway. They also articulated a great deal of personal blame for losing their jobs during the dot-com crash, as we will see in chapter 5, saying that they should have been smarter, faster, or even more driven to protect their careers.

Comparison of Strategies

Table 3.1 compares the three strategies I found among Silicon Alley workers. In this typology, different orientations toward a project, a company, or a

Table 3.1
Comparison of strategies to manage uncertainty

Strategy	Financial	Creative	Actuarial
Description	"Taking a chance"	"Nothing to lose"	"Don't count on anything"; "It's up to you"
"Risk is . . ."	An opportunity	Fun; Not a big deal	Avoidable
Affiliation	Company	Projects and the product	Career
Risk is understood as	Financial portfolios	Reputation	Safety and longevity
Dimensions	Faith in the company	"Accidental big step up"	Problem solving and careful research
	Portfolio diversification	Multiple media Projects and products	Opportunities without naïveté
	Accounts of entrepreneurial value		Keeping options open
	Respondents		
Number (percentage)	13 (25%)	22 (40%)	19 (35%)
Fields	Founder	Content	Founder
	Technical		Technical
	Content		Content
	Finance/Business		Business
Background	Liberal arts	Liberal arts	Liberal arts
	Technical	Fine arts	Fine arts
	Business		Technical
Entry into Silicon Alley	Relatively late	Relatively early	Mixed
Exit strategy	Diverse portfolio of jobs, skills, contacts will mitigate failure and allow rebound; search for "next big thing" leads to new ventures	Friends, projects and reputation will help; try another city or another medium	Build transportable skill set; ensure employability; take safe haven; deal with uncertainty to remain in industry

career influenced the ways in which people within Silicon Alley framed the uncertainty they faced. This table highlights the trends in the ethnographic data and is meant to serve as a summary of ideal types of strategies. In practice, the prevalence and expression of these strategies changed over the course of the history of Silicon Alley; for example, a close devotion to creating visible, cutting-edge projects was not a career strategy that was highly rewarded by the industry during the economic downturn.

These strategies provided a compelling picture of how individuals perceive risk and uncertainty in their careers, how they hedge and insure against it, and how their risk taking relates to the requirements of an emerging innovative industry. Some worked in the small start-up firms that formed the popular image of what an Internet company should be. Others worked for corporate entertainment and media conglomerates as these companies were just beginning to explore their strategies for Internet content. And, although I did not know it at the time, one medium-sized firm in which I conducted interviews announced its initial public stock offering days after I was casually hosted in its office. What linked these workers is their sense that a change was happening, and that they felt as if they weren't just a part of this change but the forces behind it. For them, the Internet industry opened up a possibility of being simultaneously creative and professional. For all, taking a chance in a new industry seemed a perfectly natural thing to do, even though what was motivating them to take those chances—lifestyle, money, creation—differed among them.

In light of the stock market crash, one can see that the "hype" in the promise of stock market rewards in Silicon Alley, as well as in other high-tech job markets. As the interview data in this chapter show, people welcomed risk not necessarily because of the expected monetary payoffs, but because they thought they had to accept risk along with autonomy, creativity, or a "good job." Other researchers have written about the entrepreneurial labor required of those working in Silicon Alley, and the destabilized hierarchies prevalent in this and other media industries that give rise to worker's sense of ownership and power over tasks have not been able to explain the attraction of risk and uncertainty among this workforce.[12] Of course, for the quarter of the workers attracted by financial rewards, these flourishes were the main reason they were attracted to the industry. Those who felt they were "taking a chance" within the industry were lured by the possibility of rewards. This type of entrepreneurialism at work ties financial performance to employment decisions in a calculative gamble in what Frank and Cook have called a "winner-take-all" market.[13] But for others in Silicon Alley using the creative approach to managing risk, the

Internet industry represented a possibility of more—not less—stable work than that offered in traditional "old" media, as well as the possibility for "creative," "interesting," and "fun" work. These were, after all, "cool jobs in hot industries."[14] Still others using the accounting strategy felt themselves capable of outsmarting risk by carefully hedging their venture labor investments in the field. These cultural frameworks for evaluating risk encouraged an orientation toward venture labor, even from people with vastly different—and often competing—ways of evaluating their jobs. These frameworks guided how people perceived and reacted to risks and how they formed rationalizations to explain the risks they were facing and make sense of their world. These cultural frameworks shaped the strategies that people used for risk and were part and parcel of the social construction of economic and financial risk within the industry. These cultural frameworks helped people working within the industry calculate the opportunity costs of the choices they faced.

These frameworks were culturally individuated approaches to risk, flexible enough to support the needs of the industry. These frameworks support the powerful ideology that the economic risks people face are the result of their own personal choices—perhaps more so from the actuarial strategy than from others, but there is still the notion that individuals must engage with and manage risk in their jobs. Still, people could and did differentiate their own decision-making process as somehow culturally authentic compared to people working with different decision-making strategies.

These evaluative principles are not particularly new, but in light of the changes to organizational forms they are newly important and substitute to some extent for other types of attachments to work. Rosabeth Moss Kanter, for example, framed three logics of careers: professional, based on the logic of developing specialist occupational skills; bureaucratic, based on the logic of advancement within a corporation; and entrepreneurial, based on growth occurring "through the creation of new value or new organizational capacity."[15] To some extent everyone working in Silicon Alley worked under what Kanter would describe as entrepreneurial career logics. But whereas she gives examples of these career logics as residing within different jobs and different companies, the strategies of financial, creative, and actuarial approaches to risk could be found within the same companies and across types of jobs. Ross wrote that for "no-collar" work the "intangible rewards—recognition, stimulation, responsibility—offered by the jobs are almost as important as the final compensation."[16] These strategies both confirm and contradict this. Far from "intangible rewards," creative risks are, for some, the driving motivation for taking particular risks and show an entangled engagement with the principles of capitalism even from people who critique them.

Just when workers across the postindustrial economy suffered from a lack of commitment from their workplaces, dot-commers felt a part of a growing industry, and were often invested—whether financially, emotionally, or both—to the companies they worked for, to the projects they were involved in, or to a career in a new industry. For some, this commitment was expressed in the actual investment of capital, through owning stocks, or having options to buy stocks, or through initial financing from themselves, their families, and their friends. Work in the industry inculcated a deep sense of entrepreneurialism and risk taking among employees that factors into what Batt and her colleagues have termed the "counterintuitive career," a career progression shaped by personal networks that leads to more independent and entrepreneurial jobs rather than to stable full-time employment.[17]

In this way, these strategies are coconstitutive of the industrial space of Silicon Alley. Workers individualized uncertainty in ways that conform to their own conceptions of the worth of their work, and these conceptions were utilized by the industry in order to make the acceptance of uncertainty more palatable to individual workers. This is what Nigel Thrift has described as a "material-rhetorical flourish intended to produce continuous asset price inflation."[18] That rhetorical "flourish" had real implications for the upper reaches of a workforce entering into a postindustrial labor market. If, as Thrift has written, the culture of the new economy was a "performative legitimation, a realignment of knowledge and power which would take in and work with middle class management bodies and desires," then certainly the rise of entrepreneurialism at work is a crucial aspect of that legitimization.[19] People working in the Internet industry used various strategies to frame the risks they faced in light of a tumultuous ride through the market cycles of information technology. Popular media highlighted one aspect of these cycles: young technology industry millionaires struggling with "Sudden Wealth Syndrome," but not the artists-cum-waitresses, poorly paid book editors, or freelance writers who jumped at the chance to work in the jobs that seemed *more* stable and more professional than then ones that they left. Stability within this industry was relative, and some of the creative risk takers found that their jobs were more secure to them than financial risk takers moving from so-called boring Wall Street jobs.

How these strategies get employed to manage uncertainty, who uses which one, and when a particular strategy becomes relevant is the story of Silicon Alley's labor markets. These strategies highlight the different approaches to economic uncertainty undertaken by people who worked within the industry. Calculating value is a mechanism for dealing with risk, and this processes forms the basis for what Susan Christopherson has called the strategies and institutions that emerge as intermediaries in high-risk, project-oriented work environments.[20] People working in this field were exposed to the risk and uncertainty that Vicki Smith found had crossed the "great divide" that previously separated low-end and high-end jobs.[21] Information technologies and collaborative and distributed work environments play a special role in the rise of entrepreneurialism in new media work, and the rhetoric of the new economy fueled a belief in the possibility of wealth-sharing at work. The phenomenon of new media workers taking on uncertainty—whether by choice or not—points to a larger social trend that has unfortunately placed more of the American workforce at risk.

The strategies to maintain employability evidenced in many of these interviews were pursued to buffer risk stemming from a particular company. Job-hopping, networking, lateral ties, and keeping recognizable companies on one's résumé are strategies to balance the risk of the industry, these respondents reported. These strategies work when the network has jobs to distribute or the industry is still around to recognize the "wow" factor built onto a résumé page. In each of these cases, there is an element of workers' keeping one eye on the market, on company performance, or on product reception as they build their career portfolio of positions. Even the less entrepreneurial workers reported that they thought others were "naïve" for not following the market trends or that they themselves, like Lucy, calculated the values of good creative work versus the value of a good-paying job. One former Silicon Valley employee and industry observer commented that what was at stake in the dot-com bomb was "people gambling for autonomy."[22] What comes through clearly in these two situations of uncertainty is that it became transformed into the price that must be paid for autonomy.

At first reading, the equation of job and investment that Alan and Sam make seems to prove their own concerns for stock option wealth. Yet, put into context of their strategies for employability, seeking good "investments" only makes sense in a labor market transformed by market evaluations of worth. This, too, is another way to outsmart the market by having a diverse portfolio of jobs that might attract the next employer or may provide the skills for the next big opportunity. While Dan, the founder, had a difficult time reconciling *values* of a company with the plummeting market *value* of a company, he too thought that these multiple mecha-nisms of calculating worth could pay off if framed in terms of skills and experience for the next job.

How people framed these risks was determined in part by what they saw as the opportunity costs of working in Silicon Alley and by the social forces that encouraged them to incorporate entrepreneurial practices into their own work. These practices show that people are adapting to a new, generalized work insecurity. Employees' sense of risk is balanced by an attraction to entrepreneurialism and to desirable cultural aspects of the work, as well as by the genuine feeling of "being a part of" their companies and by the possibility of rewards. In this way, just as venture capital was used to invest in the industry, so too were the hours of labor of people working in the industry. Even though people using the financial strategy were the most explicitly entrepreneurial in framing their career choices, everyone demonstrated some kind of venture labor—of using their labor time to invest in projects, ideas, companies, and their careers. Commit-ment to a particular project or company or to an individual's career stabil-ity often coincided with companies' goals, even if the connection between workers' goals and their companies' seemed at first to be at odds. How workers tried—unsuccessfully—to hedge against those risks is addressed in the next chapter.

4 Why Networks Failed

When Silicon Alley began to reel from the effects of the dot-com stock crash, one of its central figures called for people to show how they were managing. Bernardo Joselevich ran a small dot-com business and published a weekly email newsletter about networking in Silicon Alley. His call for strength was not for more advertising, or financing, or for lectures on best practices. It was instead for people to go out to parties. He wrote, "What a privilege to be in a place and time where you can prove your resilience simply by showing up at parties.[1] Later that month he reminded his readers that "partying and schmoozing are not about being in the mood. It's not necessary to have something to celebrate, it's a business activity, and not the most unpleasant one."[2] Of course, Joselevich was a bit biased: his weekly email newsletter listed events including parties that were happening in Silicon Alley or of possible interest to the people who worked there. At its height, the newsletter listed over fifty events each week. But Bernardo's writing about "partying" and schmoozing as a serious business function shows a shift in discourse toward the *network* as a source of power.

This chapter looks at the success and failure of the work that happened at these parties in terms of the impact on people's careers and job security. Parties in Silicon Alley often linked club culture and business culture, a feature that Angela McRobbie has noted is common in the labor markets for youth-oriented cultural industries.[3] The dense networks of Silicon Alley helped foster innovation and build the industry but ultimately failed to support the people who made those connections in the first place.

The seemingly carefree, hedonistic image of successful young people enjoying nightlife in Silicon Alley was a carefully crafted one, which "could be as articulate in expressing a company's profile and aspirations as the corporate portfolio," as Andrew Ross wrote, and companies took seriously their image as projected by these events. Since Courtney Pulitzer, Bernardo Joselevich, and several other trade media outlets covered such events and an informal network of buzz about them, parties became another way for companies to communicate their position within the industry. A party thrown by a dot-com had "significance more akin to a high ceremony of state" in which "all of the decisions that go into the event are closely scanned and interpreted by employees and onlookers as if they were intended to send a formal message about the company's current standing in the business world," according to Ross.[4] Razorfish, the company studied by Ross in *No-Collar*, set a standard by which other companies parties were judged. At its annual May Day party in 1999, the company celebrated its successful IPO with "a vertically-challenged, bubbly clown-girl popping out of her outfit" who handed guests oversized lollipops, an inflatable kids "moonwalk" where people "gyrated (in various levels of undress and various levels of intensity) and smoked (various types of plants)," several rooms for "making out with a sweet one, or a strange one—of which there were plenty," a real tattoo parlor, and "young ladies in underwear who were dancing on the stage."[5] Icon, the company that published the pioneering online magazine *Word* threw a party in Audart Gallery on Broad Street in the financial district in an abandoned Swiss bank. Even though the zine had such a high culture pedigree, the event featured "scantily clad go-go girls dancing on raised cubes to cheesy mid-sixties fake pop music in the front room." To which, Courtney Pulitzer wrote, making a reference to the tacit, unspoken competition among Silicon Alley companies to throw the best, wildest, most outrageous party, "Take that Razorfish!"[6] Pulitzer also likened a smoky, rave-like loft party sponsored by online music company SonicNet (later sold to MTV) to "Razorfish on steroids."[7]

Josh Harris, founder of the start-up company Pseudo.com, threw infamous parties in his 10,000-square-foot live-work loft in SoHo that included go-go dancers, live performances, and boxing "operas" along with computer games, live video art, and robotics, like the one shown in figure 4.1 (figure 4.1). Both by Harris and by the media that covered him, the loft and its denizens were likened to Andy Warhol's Factory. These "happenings" included "club kids, techno-music, drugs, models, cross dressers and computers" and the "geeks who would later become the CEOs of Silicon Alley companies."[8] Even after millions were invested in the company, Pseudo continued its wild and often endless office parties. The company held raves with art installations of naked women, carnival acts, and outrageous costumes including a sea creature, a dragon, and a woman who was naked except for the credit cards strategically covering parts of her body, and in 1999 it sponsored a continuous month-long millennium celebration party.[9]

Figure 4.1
A party at Psuedo.com

Raucous and risqué behavior at such events was widespread across Silicon Alley. Even a post-IPO company deliberating trying to cultivate a professional image hosted a party with professional dancers silhouetted on a scrim, and the CEO of venture-backed company was seen wearing leather pants at his company's party in a nightclub. But these parties weren't limited to young, edgy downtown start-ups. Banks, law firms, executive search firms all hosted events, like the one that executive search firm Redwood Partners threw at the Angel Orensanz cultural center in downtown Manhattan where staff from the New York City mayor's office mixed with people start-ups and venture capital firms.[10]

However, as Elizabeth Currid argued for other cultural industries, it was not that "creative people hung out in these places and got drunk, snorted coke, and danced all night" but rather that these were "also the sites of meaningful social and economic interaction" forcing "us to look at entertainment venues in a totally different way."[11] For companies in the start-up culture of the late 1990s and early 2000, the industry party circuit served as a way to advertise, announce successes, and continue to build networks and support for their products and services.

Currid argued in *The Warhol Economy* that New York cultural and creative industries thrive on a density of nightlife. However, little has been written about how these events actually *work* for the people who attend them. How important was networking to people working in Silicon Alley? There emerged a logic that networking was good for networking's sake, that making connections would have some sort of payoff even if those could not be foreseen at the time.

Social Networking's Cultural Moment

Networking, within the common business parlance, refers to the meeting of potential business-related contacts in a setting outside of work. Within Silicon Alley, workers thought of networking as a crucial component of their jobs, whether for their own career advancement or for the visibility and success of their companies. Within Silicon Alley there were explicit times and places, both online and physical, for networking including parties, conferences, tradeshows, meetings and lectures. Networking through email lists, online newsletters, and other online communities provided another venue for those working in Silicon Alley to meet one another, and in the early years these were key sources of legitimacy and validation within the industry. Networks are one form of support for venture labor, but the history of Silicon Alley shows that social networks as employment support are ultimately problematic.

Networks of People and Technology

Why did social networks and business networking become so important during the dot-com boom? This isn't to say that social networking is new, but that as a concept, theory, and practice networking in the Internet industry took on an importance bordering on frenzy. While the party culture of Silicon Alley may be an extreme example of dot-com networking, research in Dallas and San Francisco technology communities suggests that the phenomenon of networking events was widespread.[12] The prevalence of "networking" ideas in the media also reflects an increasing importance within popular culture. In the popular imagination, the rise of the Internet linked computer networks with social networks.

Yochai Benkler argues that individual actions, and especially linked, cooperative ones that do not depend on market forces, are playing a much greater role now than they did in the "industrial information economy."[13] Benkler's argument rests on the notion that communication networks and social networks are rising in tandem with one another—that the power and capacity to link technically is supporting increased social connections, or as he phrases it "the increased capacities of individuals" with a wider range and greater diversity of social ties are "the core driving social force behind the networked information economy."[14]

During the explosive growth of the commercial World Wide Web—both in terms of the numbers of users and in the content that was being created for it—computational metaphors and models were linked to social ties. Social theories of connection tied these together. Social networking as a practice found its cultural moment and its technological affordances in the technical networks on the Internet. Manuel Castells's theory of the "network society," which first came out in 1996, also referenced the idea of social network changes tied to the rise of computer networks—that emerging computer networks support, foster, or reinforce social ties among people.[15] Within the popular press examples, the Six Degrees of Kevin Bacon and other social networking games became popular, and business and career advice columns advised things such as making "networking a career habit."[16] By 1998, an industry survey of job hunters reported that 61 percent of respondents found jobs through "networking."[17] Malcolm Gladwell popularized social network theory in his article on Lois Weisberg in *The New Yorker* magazine in early 1999 and later in his best-selling book *The Tipping Point*.[18]

Within Silicon Alley, the workers whom I interviewed took for granted that they had to hustle for their next projects and jobs. Attending events in Silicon Alley and participating in online communities were considered by many to be prerequisites for continued employment in the field. Going out to the numerous events was "work" for most, and few respondents reported enjoying the cocktail parties and nightclub events that accompanied work in the field. This privatization of job security—through individuals' net-working activities—means that building the social capital of an industry is seen as the work of individuals, not as a function of the industry as a whole.

People in Silicon Alley saw parties as one of their most important business activities and spoke of their social connections within the industry as a form of unemployment insurance, their hedges for risky ventures, and buffers for difficult times. Friends and acquaintances in Silicon Alley were often thought of as the fastest and most reliable sources for information about jobs, rapidly changing technologies, trends, and prospects for clients or projects. For those working in Silicon Alley, social networks like these were seen as critical in maintaining their skills, knowledge, and employ-ability. And those social networks were formed in the places where people hung out, either virtually or literally in dense geographic clusters. Social networking and the networks that people build in their industries, I argue, are needed for building a flexible workforce and sharing in the economic uncertainty of the industry. The history of social networks in Silicon Alley teaches a cautionary tale about what happens to social networks during difficult times such as that which occurred during the dot-com crash in 2001.

On the one hand, the social networks that developed in Silicon Alley built a community that supported small companies and helped develop a regional presence for a young and growing industry. These networks were also critical for employees' own sense of job mobility. However, a tension developed between how social networks supported individuals' careers and how companies benefited from these networks. Social networks helped manage and shift the risk of the industry, as is often the case in creative industries. Social networks became buffers that helped people manage the risk of one job with the prospect of quickly finding another. [19] These buffers failed in the economic downturn. Social networking is certainly not unique to the Internet industry. Other industries—and media and cultural industries in particular—also rely heavily on the networks of those working within the industry. Professional and middle-class workers readily report increased pressures to "network," or make potential business contacts through social engagements. Nor are the structures that emerge from social networking new—certainly the "old boy's club" metaphor and the image of the three-martini lunch predate post-Fordist changes in employment structure that encourage workers' reliance on network resources in the face of shrinking organizational supports. For example, Charles Kadushin identified "lunch distance" as a force that concentrated the American intelligentsia in a radius around midtown Manhattan from which a writer could reasonably travel for a lunch with his or her editor.[20] And entrepreneurial companies have dense informational, legal, and financial ties within their industries and among the companies that support them. There is less written, however, about the kind of project-based and temporary organization among the workers in such creative firms that Gernot Grabher has called a pool of resources that "'gels' into latent networks."[21] Several scholars have suggested that networks increase workers' mobility within industries that rely on network forms of organization, and regional networks may substitute for types of workforce support that used to be found within organizations, such as internal labor markets, job training, and job security. How workers fare within regional networks is still unclear. Scholars of technology industries have noted that regional networks provide many resources for workers as well as for organizations, but may also prevent workers' mobility across regions. Social networks provided information on local markets and served as a new form of "labor market mediation for workers," supplying workers with a type of job security in which personal connections serve as conduits for information about new jobs and new technologies.[22] In Silicon Alley, the absence of other organizational and industrial supports meant your social network became the main resource for maintaining employability.

Within Silicon Alley, being good at networking was recognized as being important to success within the industry. In their first ranking of the Silicon Alley's top 100 movers and shakers, *Silicon Alley Reporter* ranked entrepreneurs and others in the industry by their vision and execution, fundraising, and a metric called simply "network": "Our final rating, network, is perhaps the most important. The most successful companies in Silicon Alley, and on the Internet as a whole, are those that are able to partner and collaborate (you know, that whole rising tide raises all ships thing)."[23] This "rising tide" reflects the idea common throughout Silicon Alley that regional growth was good for all (or at least most) individual companies and employees. This growth was thought to be generated, in part, through networked reputations. The practice of networking provided ways in which to build companies, reputations, and resources—but it was also a cultural concept that arose at a particular moment in history. That moment coincided with the rise of postmodern ways of working and the rise of venture labor.

Building Communities, Circles, and Industries

Within any industry, there are several ways to create a sense of community among the people and companies in the same field, and in many ways Silicon Alley was not exceptional. Early on several associations formed to build community across the new industry. Competing trade publications covered the goings-on of local businesses. Daily newspapers and general interest local magazines reported on a growing new industry and its rising stars. These are all common across many different types of industries (including my own, academia) and across different regions. However, people working in Silicon Alley along with those who covered it focused on creating community with an enthusiasm that bordered on frenzy.

The New York New Media Association (NYNMA) was one of the first Silicon Alley industry associations. Founded in 1994 with a $50,000 grant from the New York Economic Development Council, its mission was to serve the needs of a growing industry even though what that industry did was not yet clearly defined. It was cofounded by Mark Stahlman, who had been an investment banker at Alex.Brown and a financial and strategic advisor for media and technology companies. Stahlman's vision of the New York new media industry was that it could be a creative and lucrative community: "The Microsofts of the late 1990s are likely to be people working in small creative teams who are thinking about inventing new businesses and models and new industries. That's the promise of new media, and that's why New York has a strong claim on being an important geographic center."[24]

NYNMA boasted over eight thousand members at its height at the end of 1999 and a professional staff of fourteen.[25] The organization lobbied state policy makers, created a visible organizational face for the emerging industry, and published economic surveys and data on Silicon Alley. It also sponsored regular workshops, lectures, panel discussions, and meetings with venture capitalists, as well as SuperCyberSuds, the largest Silicon Alley networking event.

The World Wide Web Artists Consortium (or WWWAC, affectionately pronounced "wack," slang for cool) was another industry association, albeit one focused less on business development and more on individual members, many of whom were Silicon Alley's more creative and edgier employees. WWWAC was created to bring together people working in a wide range of companies: designers of interactive kiosks, copywriters for corporate communications who were exploring the new media of the World Wide Web for company brochures, and CD-ROM programmers who were dabbling in html, the coding language for Internet web pages. WWWAC was founded in December 1994 and sponsored smaller-scale "special interest group" gatherings to encourage people to share ideas about particular topics of interest. WWWAC founders used a combination of online and offline connections including email lists, parties, picnics, and panel discussions. While the young Internet industry in New York may have resembled many other industries in this way, it differed in the extent to which young people were involved, in the number of events held, and, as we'll see in this chapter, in the nature of those events.

Silicon Alley had another key difference to established industries in that online groups were crucial for defining their own sense of community. Especially in the early days of Silicon Alley, those who said they "got it" were connected through online news or online email lists, and these relationships translated into the networking practices of the Internet business community. The online Bulletin Board Service ECHO was a central hub in the creative networks of Silicon Alley. Founder Stacy Horn explicitly modeled ECHO after the West Coast's influential service, WELL, founded by Stewart Brand.[26] But these networks were explicitly geographically based as the "East Coast" in the name suggests. As Horn said, "The online connection isn't complete until you meet someone off-line."[27]

Trade publications formed another sort of community. The two biggest for Silicon Alley were the online newsletter *AtNewYork* and *Silicon Alley Reporter*, which published a print publication. Both provided a place for industry-specific news to be reported and community views to be discussed. More important, they created a sense of community through reporting on local businesses and events.

Founded by Tom Watson and Jason Chervokas in 1995, *AtNewYork* began as a weekly e-mail newsletter about New York's new media industry.[28] Calling itself "Silicon Alley's hometown newsletter," *AtNewYork* combined a sense of pride in its journalistic integrity with a reflexive recognition of its role in building Silicon Alley, a community in which the editors often included themselves.[29] When Chervokas departed in May 2000, he described the newsletter's mission as being to "report on the first homegrown start-up industry in New York in 60 years, to experiment with e-mail publishing, to deflate the hype created by the tech industry PR cycle that feeds reporters into the hands of 'analysts' who are often paid consultants for the companies that reporters are covering, and to serve as a sort of community newspaper for Silicon Alley."[30]

Along with creating a sense of community in Silicon Alley, publications like *Silicon Alley Reporter, AtNewYork, Alley Cat News*, and *ArtByte* helped define what and who were New York's new media companies. They did this by framing the coverage of issues, ideas, technologies, companies, and personalities that made up the growing industry. In many ways, these publications functioned as what sociologist Charles Kadushin has called a social circle. Social circles are a model of industrial organization distinct from social networks. In networks, the idea is that people are tied or linked to one another. Social circles are more diffuse and in a way more powerful. As a conceptual model, social circles show that insiders can recognize who is legitimately in or outside of a particular scene or field, without necessarily knowing them or being connected to them by one or two degrees— they don't need to be friends or acquaintances to be recognized as legitimate insiders. Social circles are how people recognize legitimacy within an industry. Sociologist Howard Becker's term *art worlds* is similar in that it describes all those involved in the production, distribution, reception, critique, and reproduction of a cultural good. The ways in which people talked about Silicon Alley as a community reflects this notion of production by social circles or art worlds.[31]

For example, an early cover story of *New York Magazine* crowned the "Early True Believers" of Silicon Alley, a highly visible group of community insiders, including Mark Stahlman, the founder of NYNMA, and Jaime Levy, the founder of Electronic Hollywood and hostess of the "Cyberslacker" parties.[32] *New York Magazine* featured the "New York Cyber Sixty" in 1995; *Silicon Alley Reporter* had the annual "Silicon Alley 100." Such press coverage from within and outside of the industry generated attention and cemented a core and outer periphery—in terms of visible companies and people—within the field. The model of social circles reflects the importance of generalized recognition, "buzz," and industry reputation—not just an exchange information or direct and indirect links as is theorized with social network theory and analysis. So links are not about transmission of goods or the enactment of social capital, but rather the creation of what we might call a community, with members learning to *recognize* one another even if they don't know one another. In the process they define what the industry is, a process that is exceedingly important for the production of artistic, creative, and cultural goods and services.

Such coverage also helped define what kinds of activities constituted Silicon Alley. For example, *New York Magazine's* "Cyber Sixty" included a successful magazine designer turned media services designer, a programmer turned game designer, educators with programs at New York's universities, a technology publicist, advertising executives at advertising firms Saatchi & Saatchi and Poppe Tyson, executives at magazine publishing houses like Hearst and at entertainment companies like Sony, independent CD-ROM producers, and digital artists and computer animators.[33] This wide range of business types and activities represented the diversity of activities in early Silicon Alley and an openness in the industry's early stage over its definition and potential directions.

The Importance of the Schmooze

Schmoozing in Silicon Alley was pursued with vengeance. During the dot-com boom there were companies dedicated to helping plan corporate parties, several people regularly reported on parties, and specialized event spaces flourished in the districts where Silicon Alley firms were located.

One of the earliest features of *AtNewYork* was a biweekly column reporting on Silicon Alley events. Calling herself the "*AtNY* gossip columnist,"

Jennifer Pirtle reported a wide array of events that included art exhibits at mainstream museums such as Cooper-Hewitt National Design Museum and the Museum of the Moving Image. The coverage also extended to nightclub parties with technological themes such as the regular "Click + Drag" party, which was a "far-out yet decidedly unpretentious 'happening' featuring drag queens, cyberculture, Internet . . . , S/M and fetish themes, and live performances," and "Rhizome@Robots," which featured top DJs and projections of film, computer graphics, and computer animations by digital artists and which founder Mark Tribe described as "highly charged with creative energy."[34] The column also reported on Silicon Alley–related conferences and panels and web site launch parties along with private events such as the birthday parties or going-away parties of Silicon Alley notables. When she left her position in April 1997, Pirtle wrote, "Who would have thought that one industry could sustain so many events!"[35]

The industry would soon sustain many more. Courtney Pulitzer succeeded Pirtle as the social columnist for *AtNewYork*, and the column became a weekly feature. Pulitzer came to *AtNewYork* from SmartMoney Interactive and Young and Rubicam's New Technology department. More important, she had been a party organizer for the World Wide Web Artists Consortium.

Silicon Alley events were covered in "The Cyber Scene," "Bernardo's List," *Silicon Alley Reporter*, and *Alley Cat News* as well as in general media coverage such as *New York* magazine. From 1996 to 2002, "The Cyber Scene" covered over 900 Silicon Alley events and listed over 8,500 participants in those events. The reporting in "The Cyber Scene" extended to the most important gatherings, such as meetings of the local Internet industry business associations, as well as more intimate events. Reported events included educational forums (such as conferences, panels, seminars, and workshops); private social events of people working in Silicon Alley (going-away dinners, engagement parties, and the like); and public social and networking events, such as award ceremonies, company launch parties, and regular networking gatherings. Compared to the number of events announced on "Bernardo's List," "The Cyber Scene" reported on roughly 20 percent of the publicly announced events in Silicon Alley, for an average of 4.6 events *per week* over the 1999–2002 period, which is, of course, roughly one event per weeknight (table 4.1). "The Cyber Scene" is far from a random sample of social events in Silicon Alley—after all, it represents the reporting of mainly one person, Courtney Pulitzer, who wrote most of the articles from 1997 to 2002. It is also one of the richest sources of historical material that exists to date on the Silicon Alley community. Even

Table 4.1
Summary of cyberscene coverage, 1996–2002

Year	Events	Number of people	People per event
1996	29	125	4.3
1997	113	645	5.7
1998	135	867	6.4
1999	211	2,399	11.4
2000	236	2,255	9.6
2001	150	1,416	9.4
2002	72	752	10.4

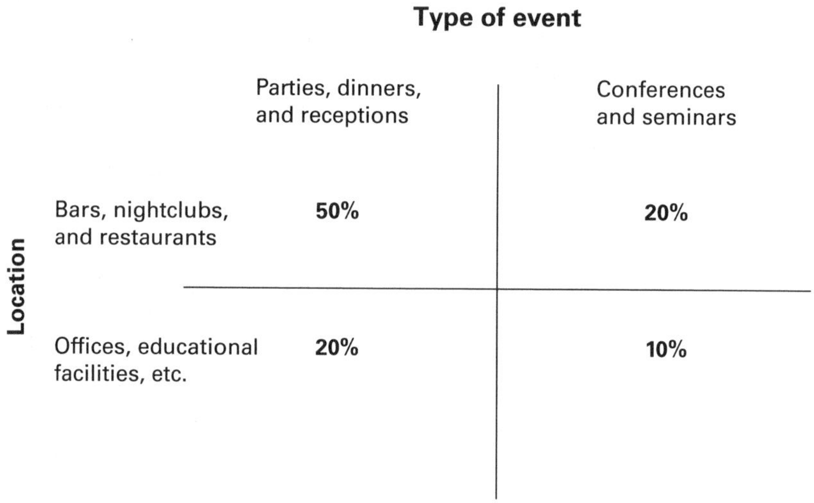

Type of event

	Parties, dinners, and receptions	Conferences and seminars
Bars, nightclubs, and restaurants	50%	20%
Offices, educational facilities, etc.	20%	10%

(Location)

Figure 4.2
Types and locations of Silicon Alley events

if the Pulitzer data is not representative, it is telling: she was an informed observer, and she thought the events she was covering were particularly newsworthy or interesting to her Silicon Alley-based readers.

Roughly 70 percent of the Silicon Alley events reported in "The Cyber Scene" were parties, receptions, dinners, and the like, with conferences, panels, and seminars making up 27 percent of the total number of events. Half of all events reported were parties, receptions, or dinners held in nightlife locations—bars, clubs, restaurants, or such. Conferences or meetings held in offices, universities, or conference venues comprised only 20 percent of the events (see figure 4.2). By the height of the dot-com boom,

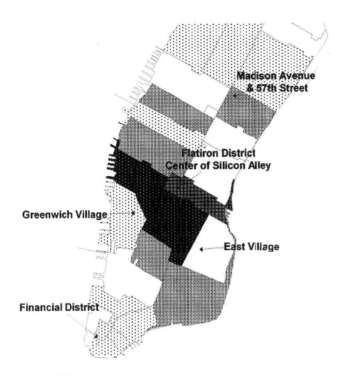

Figure 4.3
Location of nightlife venues for Silicon Alley events, 2000–2002

these events clustered densely around the Flatiron district that was home to many of Silicon Alley's companies (see figure 4.3).

The concentration of the production of Internet content in New York and the socializing frenzy that came with it reflected what Matthew Zook called the new economy's "remarkable degree of clustering despite its much ballyhooed spacelessness."[36]

While Pulitzer's columns were far from a complete listing of all events in Silicon Alley, their coverage extended to the most important industry gatherings such as NYNMA and WWWAC meetings. Panels featuring industry leaders were frequently covered, as were networking breakfast meetings, parties celebrating new offices, companies, or products, and annual awards ceremonies. In the first columns, the launch party for Total NY, a site owned jointly by American Online and the Tribune Company, was covered; less than a year and a half later, its "closing" party was also featured. Birthday parties, anniversary parties, going-away parties—in the early years of coverage, Silicon Alley, at least through the eyes of its social reporters, was a community that celebrated rituals together. Even otherwise

staid business public relations functions or informal panels became opportunities for a bit of frolicking, and both the serious and the frivolous were
found in the columns. By March 1999, events reporting had moved to
AtNewYork's web site and was no longer included in the version of the
newsletter sent by email to subscribers. Pulitzer continued writing about
social events for her own newsletter, "The Cyber Scene," and formed her
own company, Pulitzer Creations, that ran both the newsletter and regular
social networking events called "Cocktails with Courtney."

With control of her own publication, Pulitzer expanded coverage of
Silicon Alley events and number of participants named for each event she
reported increased dramatically in 1999. Rather than serve as a document
of a downtown party scene, Pulitzer's column attempted to inscribe the
diversity of Silicon Alley businesses and events. A mention of a company
became a mode of advertising to a select group of peers. For example,
Pulitzer describes the people she talked to at the NYNMA Cybersuds party
in January 1999, a large gathering held on an empty floor of an office
building, demonstrating how social events reporting was important for
company publicity: "I met Catherine Winchester, CEO of Soliloquy, Inc.,
whom I'd heard about before, but never met. She explained the nature of
her company to me. It's an interesting company in that the name indicates
speaking alone and the product is about conversing with a computer.
Standing by Oven Digital's display table, I saw Michael Hughes of Oracle,
who sent me a most interesting link the other day about Oracle and its
new $100 million venture fund to foster companies that leverage its Oracle
Internet computing platform. There's an incentive!"[37]

The mention of businesses in such a column became a kind of advertising. Companies attempting to show their cultural capital of young, cutting-
edge, avant-garde design by throwing the hippest, coolest parties could
count on buzz from events coverage and industry insiders talking about
the event. For other companies, like Oracle and Soliloquy (mentioned in
the previous quoted passage), being able to connect with people to talk
about their business success was only amplified through the trade media
coverage that came with such events.

Of course, the proliferation of events came at a cost: competition. Bernardo Joselevich started Bernardo's list, an email list with the sole function
of telling people about the Silicon Alley events of the week, and a hierarchy
of events emerged later in Silicon Alley's history, with smaller, more exclusive events replicating the early "insider-y" feel of what some called "old"
Silicon Alley. By the end of 1997, *AtNewYork* referred to the New York New
Media Association's large CyberSuds events as "roundly mocked by Silicon

Alley insiders, but it's always been the place where newcomers discover a new industry."[38] Still, the time required to stay connected within the field was enormous, and as we'll see from the interviews presented in this chapter, many people felt it was part of their job responsibility to keep abreast of these events in order to maintain their careers.

Homophilous Sectors

The key finding of Mark Granovetter's "Strength of Weak Ties," a now classic article on how job seekers found information that led them to their jobs, is that weak ties, not strong ones, are most important for information that leads to jobs. Job seekers in Granovetter's study didn't find work through their close friends, but rather it was the friends of their friends—the weak ties in their networks—who gave them the new information that led to jobs.[39] Network diversity is important because diverse networks help people get new information—information that is not duplicated by their close, strong ties, who presumably know many of the same things that they do. Fresh ideas—and some argue innovation—come from information from weak or more distance ties. This theory rests on the idea that your friends know much of what you do and think more similar to the way you do than strangers would. But people at the periphery of your social network are connected to other people and in different ways.

In the early days of Silicon Alley events, people came from a diverse array of subsectors and related industries. However, as the industry boomed, Silicon Alley events got more and more *homophilous*, or similar in terms of the kinds of businesses represented at any particular event. That means people from that e-commerce companies were hanging out with other people from e-commerce companies and people working in the arts were more likely to be at specialized art-specificevents. This in and of itself does not seem like such a bad thing until one considers the importance of diverse networks for people's career stability and innovation within an industry.

Events in early Silicon Alley brought together people from many differ-ent kinds of businesses—both inside and outside the Internet industry. Perhaps this was because a unified industry had yet to consolidate from the distinct activities that comprised it.[40] People from the arts, print media such as magazines, technology, and Internet companies most commonly frequented Silicon Alley events in 1996 and 1997 (see figure 4.4).

At those early events, people from the arts served to link groups of people together. The strength of using data generated from Courtney

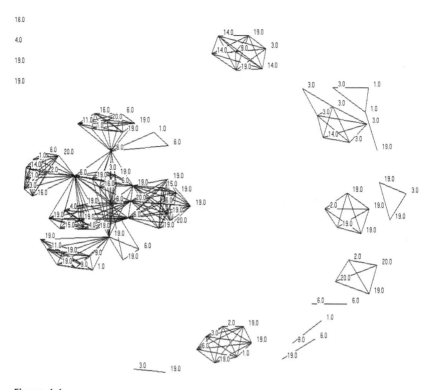

Figure 4.4
Network of individuals attending Silicon events, April–June 1997

Pulitzer's reporting for mapping Silicon Alley networks is that it provides a historical account of who was attending events together. However, this reporting is not complete, nor can it account for the other ways that people might know each other. For example, these data can show us who was at parties together, but not other people they know outside of these parties. As purposive samples of Silicon Alley nightlife taken by Pulitzer, the data from this reporting represent the kinds of social networks that formed at nightlife events, even if the data cannot exhaustively catalog all the social ties of people working then in Silicon Alley.

By 1997, dense networks had formed across these events. Linking these parties together were people who were mentioned as having been at two or more events. Most commonly, the people Pulitzer mentioned as having been at multiple events in a time period were from arts and creative sectors. That's not to say that there weren't other people in Silicon Alley attending multiple events, but rather that what we know from Pulitzer's reporting is

Figure 4.5
Network of individuals attending Silicon events, June 1999

that artists played a key linking role during the formation of Silicon Alley, connecting different parts of the industry together.

As Silicon Alley grew, so did its social networks. With growth, the parties were larger and Pulitzer began listing more of the attendees at each event. The composition of the attendees changed as well. More representatives from professional services and financial firms were reported as attending or sponsoring these events, and people working in public relations, finance, and business frequented these events almost as often as those working in Internet content companies (see figure 4.5). As the industry grew fewer artists linked the multiple events together, and the networks as a whole were dense, more homogenous (or homophilous), and less richly linked.

Early events in Silicon Alley reflected the creative and artistic image of the industry compared to the later events at which there were more people from business or finance. The networks got more homophilous over time, with people tending to go to events with people who were in sectors similar to theirs. While these social networks of people working in similar parts of the Internet industry can provide a benefit when the industry is growing, it limits the diversity ties that can be formed during such events. As a result, in the economic downturn there would have been fewer people and businesses to turn to for new ideas, job opportunities, and information.

The Work of Networking

People who worked in Silicon Alley had clear ideas that they "benefited" from attending parties and other events by meeting people who could potentially be useful in promoting themselves, their work or their companies. Almost all of the people I interviewed thought of Silicon Alley–related events as potential ways in which to advertise their companies, meet potential business partners or investors, and connect with their peers. As economic geographer Susan Christopherson phrased it, "In New York new media, who you know matters almost as much as what you know, and that, in turn, determines what kind of work you get and how steady it is."[41] Within Silicon Alley, independent workers and management alike had a keen awareness of the people who could help them get things done or help them get their next job.

While social networks may be an efficient mechanism to provide information to participants and to share technical knowledge and skill diffusion, they also functioned to distribute risk across Silicon Alley. And although there has been much written about the "regional advantages" that networks bring, few have looked at the disadvantages of such closely linked industries. While building regional networks helps a region as a whole accrue "regional advantage," there are private, and unequally distributed, gains to be garnered from the successful navigation of those networks.

People in Silicon Alley talked quite a bit about "community" and the "scene," especially in the early years of the industry. Although New York's young Internet industry was much smaller than Silicon Valley both in terms of the size of companies and in the number of employees, people in New York commonly made comparisons to what was happening in California and deliberately modeled behavior on what they perceived was the case. For example, Pulitzer wrote the following about a conversation she had during a party for *Feed* magazine with a reporter who was covering northern California's technology industry: "We were comparing what's been happening with the Silicon Alley "Scene" versus Silicon Valley. While SV has been around developing soft and hardware for much longer, it seems as if there is a stronger community sense here. He commented, parties like this one are not fringe, or trivial at all, but essential and core to the scene. It is events like this that propel the scene forward and give it validity."[42] Pulitzer's comment that parties can help "propel the scene forward and give it validity" fits squarely with theories of regional growth. Networks generated from "events like this" are "essential and core" to developing what Pulitzer called "the scene," or the regional industrial network. Or as Elizabeth Currid has argued, scenes have a "dual purpose" of promoting one cultural product while simultaneously providing the site for further cultural production.[43]

In her book *Regional Advantage*, AnnaLee Saxenian found that successful high-tech areas like Silicon Valley had established "network based industrial systems" that are "organized to adapt to continuously fast-changing markets and technologies."[44] Such regionally based networks encourage collaborative practices across and within organizations, allow for the rapid diffusion of continually changing technical information, and build environments of innovation that provide positive economic externalities for firms and workers. Saxenian and other scholars have argued that such networks support regional growth, especially in high-tech or rapidly changing industries.[45]

There are several reasons why these regionally based networks emerge in high-tech fields in general and in the Internet industry in particular. Informal communication through social networks provides a mechanism for information about new technologies to be rapidly diffused, which is more valuable for rapidly evolving high-tech industries than is information from other, slower media such as technical journals.[46] The benefits of networks to an industry also include complementarity, adoption, lock-in, and scale.[47] Complementarity refers to the technical compatibility, such as the emergence of standards or conventions of working with new technology. Benefits of adoption refers to how a networked group adopts new technologies more quickly than isolated individuals—a kind of viral form of technology adoption that we see when people working in a networked industry become avid consumers of other companies' products and services. Regional networks also help producers avoid the costs of switching, either by encouraging "lock-in" to a particular technology or by providing the resources such as information or training to reduce the costs of switching. This occurs when an informal consensus emerges within a regional network about which technologies are deployed. Another economic advantage is that regional networks can enable small producers to utilize effective scale in production through subcontracting or cooperative production relations.[48]

These regional networks provide many resources for workers as well as for organizations. They are a form of what scholars call labor market intermediation that connects jobs and workers, especially within rapidly changing industries. In addition to providing information about job opportunities, networks are a way to share information about emerging technologies, learn new skills, and keep up with rapid changes in the industry.[49]

The parties formed a sort of arms race—during the height of the boom, the notion that companies could build publicity and legitimacy through parties was prevalent. Brand awareness was seen as an avenue to company profitability, and making a circle of early and intensive technology adopters aware of the product was one means of generating buzz. This is one of the ways that the nightlife social networks benefited Silicon Alley companies. Another was ostensibly through the connections of the employees who attended such parties.

Social networks form a unique sort of resource that is at once both public or shared and private, helping support collective, industrial goals and needs while still being generated by and often benefiting individuals. Social capital, one of the kinds of resources that social networks build, is often thought of as the public use of collectively held resources. For example, James Coleman, a key theorist of social capital, conceived of it as an asset that a group can utilize when certain social network conditions are met.[50] When Pulitzer talks about building a community or scene in Silicon Alley that propels the industry forward, she's evoking this sort of definition of social capital.

However, we all know that there are differential uses of power and social capital within the same networks, and people have different capacities for building such networks and different resources for navigating them. People with families or other obligations might find the party scene difficult to keep up with. A web design firm founder quoted earlier talked about how "easy" it was to create a community when people in the industry have similar educational backgrounds and are the same age, but what does this imply for those who don't fit this model of dot-com hipster? And on the evenings when I felt awkward "circulating" in a crowded room at some launch party, I had the strange feeling of being in high school again, with the popular kids getting even more popular in these settings and the shy ones staring at their feet. British cultural studies scholar Angela McRobbie examined nightlife as both a literal and metaphorical passage point into creative industries in general, arguing that people move from "clubs to companies." Club culture patterns how people can manage their work identities, infuses the workplace with cultural norms of behavior, dictates how professionals in such industries should model their careers, and further closes gaps between leisure time and professional reputation and between creative production and cultural consumption. Metaphorically, McRobbie likens navigating the club guest list and getting a bouncer to let you inside to the informality of recruitment and hiring in creativity industries. But if nightclubs are literally hubs for networking in creative industries, "then age and domestic responsibilities define patterns of access and participation" and getting your foot in the door of an industry becomes as difficult and subject to the whims of fashion as getting past the velvet ropes.[51] The irony, as McRobbie puts it, is "that alongside the assumed openness of the network, the apparent embrace of non-hierarchical working practices . . . there are quite rigid closures and exclusions." These networks, McRobbie argues, provide individual solutions to systemic problems when creative workers rely on informal networking without having institutional or organizational supports."[52]

There are other types of privately held power from such collective social resources. Ron Burt in his theory of "structural holes" addresses the power of brokering—of literally bridging gaps in a network. People positioned at the intersection of two otherwise disconnected groups can use their position for a kind of power.[53] Within Silicon Alley, Susan Christopherson identified personal social networks as inherently exclusive because they encourage the development of "nontransparent hierarchies" that are difficult for certain groups of workers to navigate, concealing the power relations within an industry.[54] Her line of thinking suggests that the new networking frenzy that took hold in Silicon Alley and other high-tech regions duplicates many of the practices of the "old boys' network." Some people I interviewed felt uncomfortable hiring or having to manage friends, while others saw no conflict in the practice.

Christopherson also found that what made social networks such a powerful force within a region could also work against the mobility of individual employees. The predominance of personal networks within New York's Internet industry made leaving the contacts in one region or learning the ropes difficult, negatively impacting workers' national geographic mobility. In an irony of the digital world, Christopherson suggests that there may be less geographic mobility for Internet workers—not more—than other types of workers because of their reliance on these dense, regional networks for exchanging information on credentials, reputation, technology, and skills. Further, the reliance on social networking as a job-matching system creates more inequality within an industry relative to industries with other types of job-matching systems, because it privatizes into a social sphere job-market functions that were once (at least somewhat) more transparently conducted within organizations.[55]

Sid, a programmer I interviewed who had made the transition from San Francisco to New York, talked about how little mobility he felt he had, even though he had worked for big-name national clients and was known as one of "the guys that got there early" in Internet design. As he said, "People knew who I was in San Francisco. When I moved to New York, nobody knew who I was." Sid's first steps were to join the WWWAC email list: "Probably the scariest thing was that I didn't really know that many people here. A couple of years before I moved out here I discovered the WWWAC group. . . . I came out to visit a couple times, went to some WWWAC events, and that was great. That community was great. I got to know people that I know even today. They weren't in hiring positions. They were more like me, like 'Hey, this is cool. What are we going to do?'"

One of Sid's early contacts was a leader in the WWWAC circles, from whom he began to sublet office space, which meant, in his words, "I had a phone number, I had the West 25th Street address which was great [to] establish that I had my own office in the right district."

That double mark of legitimacy—the right address in the Flatiron District and a connection to an established figure—helped him eventually land his first interviews in New York. Sid saw all of these as much more important for getting a foot in Silicon Alley than he did his previous contacts, reputation, or experience. Still, he was hired by "an old friend" from California who was setting up a design firm in New York.

Regional networks replace other types of workforce support provided by organizations, unions, or associations. These include what sociologists call job ladders or internal labor markets, which are opportunities for career advancement within a company. Regional networks may also serve to help privatize job and skill training outside formal organizational settings. A study of Seattle information technology workers and the companies that employ them found that employers overwhelmingly thought of job training as the responsibility of workers, not companies. At the same time employers thought that workers who had held many jobs within the region were more likely to have higher skills than workers who had more experience within one company.[56] In fact, as shown in surveys conducted in both New York and Seattle, the perception that potential employees bring in knowledge from outside the firm is crucial for freelancers' continued employability.[57] Skills gained outside benefit the firm, raising the level of knowledge and access to new technology within. This means being able to bridge organizations is both how freelance, contract, and temporary employees get their work in the first place and a tacit expectation of what they should do once hired. Individuals, in seeking out relationships to provide for their own continued work, create ties between their company and other organizations and among industry sectors.

At first blush, there is little wrong with the picture of a rising social network tide that lifts many boats. But it leads to a kind of entrepreneurial reflexivity in which people continually monitor their own ways of *being* in these kinds of industries. Any blame for not having a job, for losing a job, for having out-of-date skills comes back onto the self. As McRobbie writes: "If we alternately consider reflexivity as a form of self-disciplining where subjects of the new enterprise culture are increasingly called upon to inspect themselves and their practices, in the absence of structures of social support (other than individualized counseling services), then reflexivity marks the space of self-responsibility, self-blame. In this sense, it is a de-politicizing, de-socializing mechanism: 'Where have I gone wrong?'"[58] Such is the work entailed in self-managing, the continual monitoring of how to "be" within the industry, a position that entails, among other things, knowing how to conduct business in nightlife settings. Building social networks as a form of support also requires this sort of individualized work. Just as one of the senior producers I interviewed said, there is a continual "performance" that must be put on in order to navigate such opportunity structures. In this way, we could say that networking has an aspect of what sociologist Arlie Hochschild termed "emotional labor"— people need to act a particular way in order to affect a veneer of dot-com professionalism.[59] It's one thing to do this as part of a job; it's quite another to be expected to keep up the performance after five o'clock.

Whether or not attending networking events led to jobs, people in Silicon Alley believed that it did. This suggests that the same the tight links among professionals that are credited with supporting work in innovative industries may reproduce inequalities between men and women, between people of different sexual orientations, between those with domestic responsibilities and those without, and between young and old workers. Certainly people living in the suburbs, or people with families to care for, would not be able to participate in Silicon Alley in the same ways. And based on the descriptions and my observations of many of the events, those unwilling to go along with the sexual innuendo, flirting, risqué behavior and displays, and public drunkenness wouldn't either.

The nightlife activities of young professionals have been credited with being an engine of economic growth not only for the industries that they are in, but also for the cities where they live. Nightlife is integral to Richard Florida's concept of the "creative class" and creative cities. Nightlife is not simply beneficial to young professionals in growing industries; it is essential, in Florida's formulation, to the innovation of those industries. Florida argues that what he calls nightlife "scenes" are important to the creative class and, by extension, to creative cities, because they offer choice, inspiration, and, in his term, "efficiency" in the use of leisure time. As imagined by Florida, in such a scene one can immerse oneself "in the bustle of sidewalks or head into an energized club and dance until dawn—or find a quite cozy spot to listen to jazz while sipping a brandy, or a coffee shop for some espresso, or retreat into a bookstore where it is quiet."[60] According to Florida, not only does nightlife inspire the creative class with innovation and creativity, it constitutes the urban spaces where they want to reside and work.

What this means is that the spatial and temporal lines between work and leisure are blurred. There is not any inherent problem with combining play and labor. However, in writings of the creative class, the work of nightlife has not been adequately addressed. Elizabeth Currid's examination of the "Warhol economy" of cultural production in New York City focuses on the role of the tight connections and the blurred lines of nightlife and business, arguing that creativity "would not exist as successfully or efficiently without its social world—the social is not the by-product—it is the decisive mechanism by which cultural products and cultural producers are generated, evaluated and sent to the market."[61] However crucial these nightlife scenes are for career advancement and cultural and creative production, they come with an unspoken set of rules and norms that could be a challenge to navigate. People working in industries that rely on this type of networking are aware of the significance of their social life to their careers but "are uncomfortable formalizing it or being overt."[62] Currid's analysis of the role of nightlife for New York City cultural production ignores the political economy of the work involved. It doesn't ask, as the urban sociologist Sharon Zukin has done, "whose city and for whom"—what kind of city such activities create and what control these activities can assert over neighborhoods and which groups of people have the cultural capital to control urban spaces through "privileged consumption."[63] While the kinds of connections that party going builds seem spontaneous and open, they can duplicate many of the pernicious old traits of the old boys' network—exclusivity, nontransparency, and inequality.

The social networks of production can tightly link firms within an industry and region together. The networks of these regional economies are particular to a specific place and are reinforced by regional contexts that shape a particular industry. However, there is a darker side to the highly linked urban economies of cultural production. There is clearly *work* involved in building and maintaining these regional economies, and in Silicon Alley this work was disproportionately done after hours by people who had the time, ability, and social capital to navigate nightlife events. Networking among the employees of an industry forms a paradox of social capital creation. Workers in Silicon Alley felt an enormous pressure to build their own social connections in order to maintain their employability; in doing so, they obtained resources for their employers and the industry as a whole, by providing visibility for their companies and through building ties among subsectors of the industry. As the downturn in Silicon Alley showed, the creation of this social capital was an effective strategy for managing the risk incurred by and at individual companies, but it did little to help buffer workers against the "systemic risk" of an industry downturn. The development of subgenres or types of companies within the industry meant people felt less affinity across Silicon Alley. The sheer numbers of types of events made it impossible to stay on top of all things digital in Silicon Alley.

But specialization came at a cost—over time, there was increasingly less diversification in the event networks of dot-commers. Of course, the event circuit was not their only source of contacts and is only a very rough proxy for what might have been going on among the eight thousand participants' own personal networks. But in terms of social gatherings, people were reporting that they needed to go out to drum up business for themselves and potential jobs, when the networks would have been well trod by others very similar to them. These events could not have possibily been a rich source of new information from outside Silicon Alley. As event networks in Silicon Alley grew more specialized, so too did the job support networks of the people who relied on these events.

One senior project manager described how he saw the line between social and business with the following: "I believe in networking, but I'm not a networking addict. I do go to industry events, exchange business cards. That's an integral part of my professional, as well social, life actually. I've met two really good friends through that networking. I sometime have networked with people who I have not benefited from directly in a professional manner and we've just become very good friends.

That's just as important. That's almost even more important. I really value that."

Even though he was not an "addict," he did go out to local events at least every other week, and even more frequently on outings with people at work, which met at least one evening a week for drinks or dinner. While it was important to his social life outside of professional development to make friendships from such events, his justification of having "networked" people who did not benefit him professionally shows an underlying expectation that networking would indeed have benefits for his career. For his next job, though, he wanted something that he could "leave at five o'clock" and for which he would ostensibly not feel the need to do so much after-hours networking.

From very early in Silicon Alley's history, networking events were often alcohol-fueled functions in bars and lofts where witty repartee and a party atmosphere made the events *seem* social even if there was an undercurrent of professional schmoozing going on. While the media covered these events as evidence of the youthful fun of Silicon Alley, many people who worked for Silicon Alley companies viewed making connections at such functions as a vital part of staying current in their careers. When asked if he goes to Silicon Alley events, a reporter for an online publication that covered Silicon Alley, vacillated on whether to call these off-hours events enjoyment or work: "I go. I schmooze. I view them as much work-related as enjoyment. Actually, they're not fun at all. You're drinking and what does the CEO want to talk about, the Mets? No! I don't think so. He wants to talk about his latest software upgrade that's going to change the face of chat. Why do I care? Talk to me between 10 and 7 on weekdays."

But, of course, he said, like many others I talked to, he felt as if he "had" to go to these functions. For some, though, networking events became their social life, and that social life could, at times, feel and be exclusive. One interactive producer described people she worked with in Silicon Alley as having "a big expectation that work is going to fulfill a social need in people's lives. And they want to socialize with people who are like them. My friend said, 'Working in new media is being able to dress like a hipster and get paid like a yuppie.'" Of course, there is a downside to this sort of expectation: as he said, people want to hang out with others who are like them, other hipsters. That made it difficult for a genuine diversity to emerge among Silicon Alley employees. A cofounder of an interactive design firm said, "There's a definite community. . . . And a lot of people involved come from very similar backgrounds as us. They're very intelligent, they're our age, they have similar sets of education, and that very easily makes a community." The drive for "community" may have excluded people who were perhaps not considered "hipsters" or were somehow different in background or education or age.

For example, one woman who was a senior writer for web sites talked about how she felt as if she had damaged her career by dropping out of the party circuit, saying, "I actually went to parties at Pseudo but not too many. . . . Then I got pregnant. That's what derailed my rise because a lot of this is about going out and networking and working a lot and I just stopped." Another woman who was a senior producer said that she was "not connected at all," which had limited the amount of publicity she got for her web sites. Her problem with Silicon Alley events was that they required "that kind of performance of 'I know a lot' [which] is fine for work, but I don't want to do that in my off hours."

Still, many people I interviewed justified how frequently they went out as a part of their jobs and saw little distinction between work-related and non-work-related events. Whether building the sense of community within the company by frequent nights out together, or building their own community of networks and contacts for future job leads, most people felt, like the two women quoted previously, that these events were crucial for keeping up with the field. For the cofounders of an interactive design company, that meant at least one night a week was devoted to "socializing at specialized new media parties." Clearly, maintaining contact with a group of colleagues, whether or not through the Silicon Alley party circuit, was critical for breaking into Silicon Alley and for getting new jobs.

One woman, a senior writer at a large, interactive media conglomerate, compared networking to looking for dates at singles events, saying that she didn't go to what she called "events events": "It's not that I have a problem with them. There's some that I did pretty regularly, but I'm pretty good at following up. . . . But you don't have to go to an event for you to do that. It just makes it a little bit easier. It's like going to a singles event. There at least, we all know what we're there for, so you don't feel like a shark digging around for cards. The whole world is a networking event, if you're a freelancer. . . . I mean, my friend is a financial planner . . . but networks on his hockey team. You know, so, to get new clients." Even though she may have preferred other types of networking, still felt as if she had to make connections and follow up on them. Nor was she alone. Eighty percent of employees within Silicon Alley reported that their social network connections were their most important source for finding jobs.

In contrast, only 11 percent reported that newspaper listings were.[64] In addition to helping people get their jobs, friends and colleagues were a key source of new skills, and ranked only behind being self-taught and on-the-job learning in importance.[65]

Another senior writer admitted that she got most of her jobs through friends and other contacts: "All the full-time jobs I've gotten in the web industry have been through a personal reference, except one where I answered an ad but I probably got the job because I knew people who could vouch for me personally. There's a strong component of 'who do you know' not 'who do you know who is important' but the assumption is that everybody knows everybody else and it's a small world and communications just flies." Later in the interview, she added that in addition to networking it was important to have a good reputation because if you're a pain in the ass people hear about that too." She, like many other people I interviewed, reported having obtained short-term jobs through people they knew through email lists such as WWWAC or ECHO, saying jobs would come "in off the WWWAC list—like [they would] say, 'I'm looking for a writer and I saw your post and I saw you used a serial comma, so you seem super competent.'" Even applying for a new advertised job was followed up with personal connections. A junior producer who was working his first job in Silicon Alley had this to say about his position: "I saw it listed somewhere either Hotjobs or NYNMA or somewhere, but in following up, I found a bunch of people that I knew there," and by the time he got to the interview he was "an employee reference" for the job.

People working in Silicon Alley felt the necessity to maintain social ties in order to stay employable. My point is not that the Internet industry is unique in the social networking practices of employees. Other industries— and media industries in particular—also rely heavily on the networks of those working within the industry. But in the absence of other organizational and industrial supports, networking in Silicon Alley became the *main* way workers sought to maintain employment security. Workers' necessity for maintaining these ties benefited organizations and the industry as a whole.

In the interview data reported in this chapter, Silicon Alley workers said that attending industry events was crucial for their continued employability. Workers sought out social networks across Silicon Alley as a type of support and for access to resources that increased their opportunities for new employment should their own companies or projects fail. More important, my research on the networking practices in Silicon Alley highlights how workers themselves experienced the "work" of networking. The quantitative data in this chapter show that there was indeed a pattern of industry affiliations at Silicon Alley networking events over time. These data suggest that particular types of companies stood to gain from the association of their employees at these events. While the work of networking was done by individuals—and the type of individuals central to Silicon Alley changed over time—benefits accrued, at least in part, to their companies.

Silicon Alley also drew upon the forms of work and organization in New York's publishing and advertising industries, which also thrive on network practices to coordinated work and information. Social events circumscribed who was in Silicon Alley. The number and types of companies reported on as having representatives at Silicon Alley events changed over time; documenting who was at Silicon Alley events helped circumscribe which companies considered themselves a part of the "community," and, in turn, the industry. Company partners could feel so out of the loop by not being on the networking circuit that they often considered themselves not part of the industry. One partner in a small firm insisted that he could not take part in my study even though his company provided the same services that Silicon Alley companies provide. He told me that he did not belong in a study of Silicon Alley, saying "we're not a Silicon Alley company—I mean, we're not a dot-com, those parties, that life, that's not us. We're just computer geeks."

Not everyone shared in the feeling that these parties were important. One junior producer who began working with an Internet company in 2000 contended such parties were "an entirely closed subculture for people on my end. There were people in sales or on the [advertising] side that got out more." Several company founders interviewed suggested that networking at Silicon Alley social events was not as important to their careers or the survival of their companies as was other, more targeted forms of networking. One founder of an e-business company had a clear understanding of the structural holes bridged by the brokers in his network, as he worked to connect to the business associates of his associates whom he meticulously researched. "Networking is crucial, although time is so precious I only go to the [events] that I think will make an impact. Going to a conference of a thousand people and paying $1,500 trying to talk to Steve Balmer [of Microsoft] or Michael Dell or [Jeff] Bezos [from Amazon]. Do they care who you are? Are they going to remember? . . . I like going to smaller events. I think spending an hour [talking one-on-one with people] is much more productive. That's my way of socializing." He also said he avoided launch parties and other Silicon Alley events, saying "those are more for social reasons" and are a "different level of networking" for a "different level" of employee.

Conclusion: The Functions of Networks

Networks became a way to manage economic uncertainty. People invested in specific kinds of networks—which, based on who was attending functions together, got even narrower as the Internet industry grew in New York. This was a great investment in employability while times were good, but this was difficult to bank on when the downturn eventually came. These networks weren't able to buffer people from the risks of the industry; rather, they were codetermined with them—in terms of social networks, as the tide rose, the tide fell. People working to stay employable create the social networks that innovative regions need to grow.

Venture labor is embedded into the process of building and maintaining networks. The work people do for companies through off-hours socializing is work; benefits companies and may become worthless in a downturn in the industry. That is because the social capital investments that people were making in Silicon Alley were *specific* to particular companies and a particular industry. An analysis of Courtney Pulitzer's reporting indicates that these networks created at industry events got more—not less—specific as the industry grew. The problem is this is that it did not help to diversify employees' venture labor. Networks allowed for an institutional rearrangement of risk and uncertainty, but the mediation of risk by through social networks was necessarily incomplete. Silicon Alley social networks were not diverse. While they worked when the industry was growing, they were not sufficiently diverse to buffer against the dot-com crash.

Networking practices individualized and privatized economic uncertainty. But networking practices were key for encouraging workers to engage in entrepreneurial labor—in the simultaneous marketing of themselves and their companies. This close proximity also produced a workforce that was highly attuned to the requirements of a networked industry. The experience of workers after the economic downturn, however, suggests that while social networks within the industry could be used to their great advantage during the economic boom, networks alone were not enough to buffer the uncertainties that individuals faced afterward. Instead, workers faced economic pressures directly, as free agents. As one cofounder of a start-up firm said, "In terms of a real sense of community, a lot of people try to foster that, but in the long term, it's still all about economics, though." The tension expressed by this respondent was clearly one of balancing the needs of the industry at large with that of the economic growth (and relatively competitive standing) of a particular company. This same tension was clearly felt by employees who were asked—implicitly or explicitly—to contribute to their companies but felt little loyalty expressed in return. Without clearly defined social and organizational structures, economic uncertainty bears down directly on individuals. These social network connections can serve as a hedge against this uncertainty for the workers who are able to access and maintain these networks, but they ultimately failed to protect them in the crash.

5 The Crash of Venture Labor

A lot of what we believed is true is not true. A lot of what we worship is not there anymore. Why is this happening? It was a bubble. We can all say that. But, why did it happen? Why did people believe in what they believed in? Why did it have such a short life?"
—Interview with Finn, a founder of a Silicon Alley start-up, April 2001

One of the main contributions of the discipline of communication is the attention to the ways in which phenomena are framed and how frames are analyzed in political and economic life. In this chapter, I examine the multiple framings of the dot-com crash and the impact of these contestations for people working in Silicon Alley and more broadly for our relationship to market structures.

In his essay on the work of French economic sociologist Michel Callon, Don Slater argues that anthropologists and economic sociologists should be studying how people calculate values within market settings, rather than denying that such calculation occurs. At the heart of his argument is the question of whether or not people are, to use economists' terms, rational economic actors and the extent to which people socially construct economic realities. Slater argued against the commonly held view in sociology and anthropology that markets are social constructs, based on perception and shaped by relativity. Instead, he argues that economic values both depend on and are at the same time *independent* of social construction. Through this process there is what Slater calls both an "interpretive moment as well as emergent objective structures." Our own cultural calculations and so-called market evaluations function seemingly objectively to us. Not only do they exist simultaneously, but they are also linked together in a way that is difficult to parse out: once put in place, market structures become real. By this, Slater means that people may have multiple, contradictory social framings on which they base calculations of value. Taken together, these multiple framings lead to a calcified, seemingly objective market reality, one that is built by the joint social intersections of many individuals' microlevel perceptions. Slater calls this the "alienation of objects detached from their networks," the intersection and cross locations of goods in multiple valuations. Once a market is put in place, it has a "social mathematics" that seems to add up independent of people's perceptions of value. In other words, according to Slater it doesn't matter what we think about economic markets once they exist, but they can't continue to exist without our belief in their "correctness" or justification of the principles of evaluation. Although Slater is specifically referring to a market for goods, I would argue the same can be said about how people frame their own positions and value within the labor market. To restate Slater, the goal of navigating contemporary labor markets is to stabilize how we fit in the market, frame our value in individual terms, and then make sense of it. This process is "a competition over the structures of markets and market relations themselves." This is what Slater means when he says that we move from calculation to alienation: that we might calculate what values we place on things, our jobs, and our work, but we become alienated in a Marxian sense when those values are decoupled from their underlying things through market processes. Social construction works to create market realities.[1]

This brings us back to the epigraph at the beginning of this chapter and the story of what happened in Silicon Alley after the dot-com crash. The young founder, whom I call Finn, was faced with the failure of his start-up, firing his friends, and months of unemployment. He captured the tensions that Slater theorized between individual perceptions and constructions of market value and the realness of it. Between people's beliefs and the financial bubble was a space of action. Sociological and culturally shaped individual perceptions helped guide decisions that people made when they, to paraphrase Finn's words, believed what was true *was* true, at least on some level Their individual cultural perceptions became calcified into an economic and social reality—an industry without a history that was strongly tied to a collection of joint beliefs in a future that ultimately had, in his words, a "short life." This is how cultural calcification happens: the layering of people's individual beliefs into a sociological reality under which decisions are made not only *seems* rational, it *becomes* rational given their sociocultural context.

The question Finn asked—"Why did people believe in what they believed in?"—becomes even more germane after the crash because the culture of risk that was so prevalent during the dot-com boom survived the dot-com crash. In this comment, he recognizes that something had become calcified, "worshipped" even, while at the same time he could not understand why this calcification was historically fragile. His struggle to make sense of the crash calls to mind a line from Karl Marx, which has been taken up by postmodern social theorists such as Marshall Berman and Zygmunt Bauman: "All that is solid melts into air, all that is holy is profaned, and man is at last compelled to face with sober senses, his real conditions of life, and his relations with his kind."[2] Or as Angela McRobbie has written, this is the way that "capitalism seems to absolve itself from responsibility by creating invisible structures, and by melting down or liquefying the old social order."[3] Why did people believe that the fledging dot-coms might have a future and how did their seemingly solid financial and creative futures melt, seemingly into air?

Using theory from economic sociology, this chapter analyzes how people represented the dot-com crash in the narratives they told about their careers. These narratives reflect a changing relationship to work in which freedom, choice, and blame collectively point toward an individuated experience of structural, economic forces. Based on these narratives, I suggest that new theories of economic value adjudicate between two positions that are most commonly juxtaposed and conceived of as in opposition, the social construction of markets versus economic rational action. Building on the work of David Stark, Michel Callon, and Don Slater, I argue that the way in which people's perceptions shape the market and the way in which people respond to the market need to be more often considered in tandem as forces of both market construction and rational action.

Framing the Crash

As with many such economic crashes, there was conflicting information and interpretations that occurred during the end of the boom in Internet stock prices. While we might now look back in hindsight over the dot-com crash and see it as inevitable, at the time few saw it that way. We now know that the Internet industry would not fully recover from the massive layouts, "shake-outs," consolidations, and closures that swept New York and the nation in March 2000. But at the time, the decline in stock market valuations of dot-com companies and company closures and layoffs were not clear indications of trends that were permanent. A dip in the NASDAQ Composite Index of stock prices in January 2000 was followed by drop of over 9 percent in value over a six-day period in March. When the NASAQ peaked on March 10, 2000, at 5,048, it was more than twice the value of the year prior. The consulting firm Challenger, Gray & Christmas estimated that between December 1999 and the end of March 2001 more than 75,500 Internet-industry jobs were eliminated.[4]

So while there may have been some skepticism within the business press about the Internet industry before the crash, the signals leading up to it were that the Internet industry was strong, growing, and would continue to grow. The idea that Silicon Alley had a promising future was at once built on many people's beliefs in the industry's growth and expansion. But once put into place, this became a structuring reality for the choices that people make. Three key cultural beliefs helped frame these choices: (1) the Internet industry had a certain, positive future; (2) existing market metrics were insufficient to measure the potential growth of the Internet industry; and (3) people had control and autonomy over their work and their choices in the industry.

Jason Chervokas wrote in *AtNewYork*, the online industry publication, that the top story for 1999 was "clearly money and the new wealth that the New York Internet industry has generated": "With Internet stocks reaching heights perilously close to the outer edge of the earth's atmosphere, everyone in New York is looking for a piece of the action. These days, if you forget your money in the morning when you go for coffee, don't worry—the guy at the deli will trade you breakfast for equity."[5] While the effects of such wealth on coffee shops might have been overrated, the amount of money flowing into Silicon Alley companies by the end of the crash was not. In the last week of 1999, over $2 billion in deals were announced by New York technology companies.[6] *AtNewYork* claimed that 1999 showed signs that the market was "catching up with some of the visionary ideas on which the Internet industry in New York was founded."[7]

Neither top financiers nor industry watchers could predict when, by how much, or for how long the markets would fall. West Coast venture capitalist Tim Draper dressed up as Batman for the March 2000 launch party of Draper Fisher Jurvetson's $100 million "Gotham" fund, only weeks before the crash.[8] *Alley Cat News* featured the headline "When the Money Keeps Rolling In: 20 VCs Fueling the Growth of Silicon Alley" on its January 2000 cover, even though that money would slow to a trickle just a few weeks after the issue ran. *AtNewYork* in its end-of-year issue wrote that money "flowed into Silicon Alley in 1999 at a pace that shocked anyone who still remembers the struggling days of shared Ethernet connections, illegal loft apartment offices, and pirated software."

In the spring of 2000, people still sought and took jobs in Silicon Alley start-up companies. Job ads posted throughout the first three months of 2000 touted the benefits of the dot-com lifestyle for the right sort of entrepreneurial employee. A company called MainXchange.com, calling itself an "incredibly rapid growth, pre-IPO teen web site," sought "several high energy individuals" and offered "Salary + options." Several job ads boasted that the company in question was "pre-IPO" with compensation such as "base, bonus, stock options, free training, pizza Fridays and cool projects."[9] And executives from "old" media companies continued to take jobs with Silicon Alley start-ups, pre- and post-IPO. For example, Xceed recruited a former chief technology officer from HBO. Psuedo.com, the Internet television company often mocked as the epitome of lavish dot-com excesses, hired a high-powered television dealmaker and executive in May 2000 away from the William Morris Agency and a former CNN executive the fall before.[10] By the time of the dot-com crash, New York City's dot-coms were not just full of twenty-somethings fresh out of college. Seasoned media executives made choices in 1999 and 2000 to join Silicon Alley companies. During the spring of 2000, MBA programs continued to offer entrepreneurship and Internet-industry-related courses at unprecedented levels.[11] The U.S. Department of Labor ranked computer-related jobs including software engineering, computer support, network administrators, and desktop publishing as the fastest-growing jobs,[12] as enrollment continued to grow at programs like New York University's Interactive Telecommunications Program, an entry point for many people interested in working in Silicon Alley. This collective social machinery of calculation and value helped create growth and stability in Silicon Alley. While it might be easy to criticize these decisions in hindsight, at the time rationally acting, savvy professionals were leaving stable jobs and career tracks for a shot at something.

The signals about recovery after the stock market crash were equally contradictory. *AtNewYork* asked just a few weeks after the beginning of the crash, "Have Alley stocks hit bottom?"[13] Companies in Silicon Alley continued to raise money. Now defunct delivery service Kozmo raised $250 million between 1997 and 2001, including $30 million in early 2001, a full year after the crash began.[14] The trade magazine *Industry Standard* asked in a headline a year after the crash if the Internet industry had reached "the beginning of the end."[15] By April 2001, an estimated 370 Internet-related businesses had closed according to Webmergers.com.[16] *Business Week* continued to herald the new economy as "reality." The dot-com crash caused a backlash in which detractors proclaimed that the new economy had been based solely on media "hype." After months of market volatility, Lou Gerstner, then the president of IBM, said at a stock analysts' meeting at the end of 2000, "Despite the hype of the media, there is no New Economy. . . . The Internet is not about the creation of new industries and new institutions." However, the notion of the new economy continued to have its strong defenders and advocates even after the crash. One highly visible proponent was Michael Mandel, the economics editor of *Business Week*, who responded after six months of stock market declines: "There is no doubt that Gerstner is right in one sense: IBM and its large customers, rather than becoming extinct, are crucial to future economic growth. But that's no reason to rewrite history. The New Economy is reality, not hype."[17] Mandel's adamant defense of the new economy plays into what Nigel Thrift called the "cultural circuit of capital," which functioned to "define what the facts consisted of and train up bodies that bent to those facts."[18] The stock market crash, just like the bubble, was a contested terrain of meaning making.

This debate over the meaning of the crash reflects concerns about what Greta Kripner has called the "financialization" of the U.S. economy, in which the role of financial profits trumps profits from productive, manufacturing, or other sources.[19] In one sense the crash highlighted the centrality of financial markets in determining valuations. Part of what is challenging is looking at these calculations within this framework, apart from judging the decisions themselves. The crash changed certain metrics of what was valuable, without necessarily changing how people framed their own valuation. In this light we can look at how people rationalized their work in light of the dot-com crash. The point of evaluating how they talk about their choices is not to ask whether these were good or bad choices, or if they accurately or rationally calculated the risks they faced. Rather, it is to extend the work of their stories to a social context and within a mechanics of calculative value. The dot-commers making value choices about their careers didn't do so necessarily on scales that economists would have recognized. Nor were they necessarily conscious of particular choices at the time. But within their narratives after the crash is evidence of career restructuring that intersects powerfully with the language of risk.

Rationalizing Failure

They way that people told stories about the crash became one way to make sense of it and their careers within it. And it is through that lens that looking at *how* they depicted their position provides evidence of the subject formation within this economic order. It is easy to ask whether it is the fault of young dot-commers to be gullible in their choices. However, before 2000, they had many signals telling them that they were working in a fundamentally different kind of economic environment. After all, even after the crash, *Business Week* could proclaim that the "New Economy [is] reality, not hype." Nor should anyone be faulted for wanting to believe that their jobs matter and the work they do is important, and, by extension, embracing market rhetoric that reflected that. Whatever schadenfreude people felt at seeing young people who had earned more money than many believed they should, their choices were "rational" to the extent that they were based on the knowledge they had, being caught up in the rhetoric of market euphoria that circulated from pundit to pundit like a raging virus.

In the interviews after the crash, several people talked about having confidence when they made choices, only to later have the terms of the choice change. As one senior producer, Eddie, said, "It's my fault in a way" when talking about being laid off from his job as a digital producer with television network "I wasn't fast, focused or driven enough. I've always wanted to have a life, and in that way it's my fault. . . I thought in a big corporation I would be buffered from the bubble, but there were other things involved." He had actively tried to be "buffered" from the bubble by moving out of small start-ups to do interactive work with a television network. He acknowledges that he thought being in a large corporation would protect him, but that he was at fault for not being driven enough or for having too much of a life. Even though the layoffs affected the entire industry—and in this case, adjacent industries such as media and advertising—he blamed himself for not working hard enough, for not being prepared enough, and for not finding a safe enough "buffer" from the bubble he felt on some level that he should have been able to predict.

Similarly, Sophie, a web producer who had worked in both Silicon Alley and Silicon Valley, showed that she had considered the risks when she said, "We thought the risk was that our company wouldn't go IPO, or maybe fail. We never thought *all* the companies would fail." Choices such as Sophie's were based on a concept that a career in the Internet industry was stable, and not particularly risky, even if work in any one particular company was. If one company failed, the experience garnered could be easily translated into another position with another firm, unless, as Sophie put it, all the companies failed. Jane, a senior writer, also used this kind of logic to defend her choices: "Working for these companies, these start-ups,

I thought they'd be fine. I had no idea that they were all going to go bankrupt, fire people and lay them off."

People tried to make sense of the choices that they had made through simultaneously evoking their own culpability *and* their vulnerability within the market. Finn talked about his own failure in these terms: "Smart people do not always win; dumb people do not always lose. Dumb people or dumb idea: could be good timing, a mediocre idea could get funding and mediocre idea would go public and those people would get rich, and there are times when smart people would not have such a fate. . . . That's how to protect the downside."

Finn thought he was one of the smart people with a good idea, so in that sense he wasn't to blame for "losing." In his statement, though, he recognizes that there is a "fate," a force larger than and outside of his own power, that determines who gets rich and which idea gets funding, and he says he used that idea to "protect the downside," or prevent him from blaming himself. He constructs a narrative of his economic experience that places him firmly in the active center of his choices, being "smart" and having "good ideas" while at the same time leaving things to "fate," or the whims of the market. Finn did blame himself for how his company folded. Finn expressed how he needed his team to continue to have "heart" if the company was going to have a chance to survive, even though he regretted having kept the company's impending failure from them: "I could have warned the team but I didn't believe it, that there was a chance that we couldn't get it and you realize how much I stress team and organization and morale. I think that is the lifeblood of a company. . . . It's about how much heart they put into it. The moment you go in and say "Guys there's a chance you'll lose your job in a month," they're not going to be that productive in the next four weeks."The lifeblood of a company is the "heart" that people put into it and that was Finn's strategy in the face of the failure of his start-up. By encouraging his team to work hard and have heart, he tried to mobilize the passions of his workers in order to beat the market. The same rhetoric of the heart and the same passions involved in venture labor that get motivated by start-ups also work to keep people invested in the project, even when the outcome looks bleak.

Steve, a senior interactive producer, talked about how other people thought they were able and willing to take risks, only to find out that they couldn't absorb the risks that they took: "The Internet really enables so many to start up their own businesses and I think a lot of people thought that they were willing to undertake that risk. We've all heard horror stories of someone who sells their house to finance their dot-com venture, which then fell on its face. . . . Actually, that's a true story, and it ripped his life apart, his family, all for this." Steve told this story to illustrate what he thought was his own agility at navigating risks and to reflect on how he absorbed and managed risk after the dot-com crash. In telling a cautionary tale, a "horror story" that "we've all heard," Steve is doing the work of both congratulating himself for not being in that position and reigning in his own ambitions and abilities in the face of market uncertainty. Steve might not have had his life ripped apart, but he reminds himself with this narrative that he could have.

While people argued that they themselves should have handled better the risk of the industry, their strongly worded narratives also pointed to the capriciousness of the market. For instance, there were people who blamed timing, "the market," and fate. Finn sums this up as follows: "[It] didn't go the way we wanted, partially because of ourselves and partially because of the market, and that's out of our control." When the market is beyond one's control, then focus on what can be controlled and how to bring oneself back into market control. When asked what went wrong with his company, Alan, a software engineer said, "The application's timing—wrong year. If it had been the year before, it would be a different story." This is in keeping with what Carrie Lane Chet found with workers whose jobs had been outsources: blaming larger economic and social forces insulated them from much self-doubt and self-blame.[20] In the process they were "accepting economic competition as both natural and desirable [legitimating] the process by which some people lose jobs and other people gain them," and were looking "to economic competition as the arbiter of their social value and moral virtue."[21]

Choice is a powerful force—people taking risks were choosing to do so, and their narratives reflect tensions between their own culpability for their situation and their vulnerability to untamable market forces. The paradox is that they then blamed themselves for the consequences of the downturn. When people talked about risk in light of the dot-com crash, they constructed an active role for themselves in shaping and directing their careers. Risk is equivalent to choice and is often framed as freedom rather than being trapped. The lure of risk—and by this I mean the idea of *taking chances*—has replaced the fear of uncertainty as the predominant economic rhetoric for the new economy. This shift was subtle but important, as *risk* and *risk taking* in economic life now imply active choices instead of economic passivity and forces beyond one's own control. Even when those forces were seemingly insurmountable, as when they were mobilized for accounts of "timing" or "the market," they were framed in such a way as to present the choices as actively made and the risks as willingly taken. People accepted and welcomed risk because taking risks offers a semblance of choice in an era when many things seem out of ordinary employees' control. The subtle discursive distinction between risk and uncertainty serves a powerful function in the new economy. The dot-com boom helped glorify risks—and shifted social and economic uncertainties to individually accounted risks.

Highly skilled, highly educated, and mobile workers were able to take on and benefit from these risks and welcomed them as an opportunity for personal challenge and growth. Their narratives reflected a feeling that they had opted in to this work, in effect choosing a riskier life, and, by extension, bringing the consequences of that risk on to themselves. The framework of risk gave them an ontological cathexis in their companies, the industry, and their work. Once that was threatened by new material realities, then their subjective position within the new economy was destabilized. How can one be a dot-commer when dot-coms fail? The individualization of risk in economic life means in part that the way in which people construct themselves as economic subjects is deeply social within the process of working as venture labor. They become tightly tied to industries, companies, and projects. But the burden of risk is something that they frame as their responsibility. Venture labor actively shapes choices and conditions through interpolation of entrepreneurial subjectivity, but is unable to access the structural causes. There are great entrepreneurial moments caused by the intersections of cultural predispositions and social movements that mobilize these values, which have the power to collectively reshape a labor market. The rise of venture labor responded to and influenced structural economic change. The transformative powers of venture labor meant that during the dot-com boom, people created new ways of working and new narratives for describing their work. This is in contrast to how Thrift and others have framed the mobilization of new economy rhetoric as solely in service to capital. During the dot-com boom people felt and were empowered within their workplaces, seeking out jobs with more autonomy, creativity, and freedom. In this way, venture labor not only served the needs of capital but also helped structure how people could have agency and under what conditions.

Did people make economically rational calculations when deciding to work in Silicon Alley? People made choices on multiple levels and based on multiple and conflicting values, and they constructed narratives around those choices. In this sense they were called to the work as willing subjects and framed their work using the cultural repertories they had been provided. The narratives that we build about how we make decisions often link the accidental and the calculative and create a story that combines feelings and passions with historical interpretation. For social scientists, figuring out that balance between the influence of culture and calculation may be the wrong question to ask in investigating how the market functions within subjective experience. As Don Slater has phrased this conundrum, "Economic sociology seems to face two impossible alternatives: either the market is absolutized as abstraction or it is dissolved into 'culture'; economic rationality approximates to neo-classical calculations or it merges with 'life.'" In order to resolve this question, he argues that we need to study the "the parameters and border disputes through which markets take endlessly contingent and unstable shapes."[22] The choices people made during the dot-com boom and the narratives they used to understand the dot-com crash is one such border dispute. While these decisions were not necessarily made consciously for economically calculative reasons, they functioned to create economic subjects within a labor market. While people are "making do" in the sense that de Certeau intended, the results of how they frame their decisions through microlevel perceptions make markets on a social level.[23]

Jane, a senior interactive producer, explained the comfort and autonomy within jobs made security seem less relevant, explaining that the cozy working environment of small Silicon Alley firms masked what later became grim economic realities. As she phrased it, "You're all family until you're not." Evoking family is an interesting concept to note here since it is precisely juxtaposed against the market and market values, involving shared values and deep, strong personal ties. Jane's description of her work environment—and by extension her choices to stay in it—was more than simply an economic relation being masked with a social one. For the work that she and many others did, these firms were like families on which identity and values were constructed. But unlike families, these ties were fleeting—you're family until you're not.

The very fabric of how this functioned—keeping relations about personal identities and shared cultural values—represents a coalescence and solidarity of the inside against the outside. It also reflects a notion that we have seen in many of the narratives that people told: that their work meant more than simply their role in the workplace. At the end, even when that failed it was noticeable and worthy of mention in the work narratives of people like Jane. Because people in Silicon Alley relied on multiple principles to evaluate their career choices—for example, financial stability or creative freedom—the narratives they told about their careers could employ values fit for the moment, narratives about what they wanted to tell at, and what worked for, particular times. This is the process that Slater calls "the alienation of objects in the form of property that can be detached from the networks in which they originate," allowing for "diverse, unpredictable and contradictory modes of calculation."[24] Dot-commers used their narratives about why the crash happened to frame and make sense of those risks and, in doing so, reified these collective values. The processes of individual meaning making, social construction of value, and seemingly objective economic values all worked simultaneously in intersecting modes of cultural calculation to build market perceptions and market realities.

The Calcification of Individual Perceptions into Structures

"Modern economies," David Stark has noted, "comprise multiple principles of evaluation"; the key is knowing which one applies when.[25] For people making decisions about their careers, these multiple principles of evaluation can be confusing. Career decisions can't be made with the luxury of hindsight, and making choices about which skills, degrees, and connections to acquire must rest on some combination of forecasting, faith, and happenchance. What was socially constructed was a collective notion that the industry had a stable future, even if work for individual companies was risky. The rhetoric around the new media "revolution" created an environment in which these choices seemed logical and rational to the people within Silicon Alley.

The frame of the financial market as simultaneously reflecting real underlying value and not being capable of reflecting true value meant that financial growth was accepted when it was working. The economic rhetoric of *Business Week* and other new economy proponents held that there were underlying fundamentals that were represented in the market, even if the magnitude of the change was greater than could be anticipated. The stock market crash brought these narratives into stark relief, making more visible the construction of the conceptual machinery of financial narratives. The narrative of the new economy intersected with another narrative that circulated about the financial market: that the phenomenal growth in the financial markets was somehow stable and predictable.

The dot-com crash was in part a clash between the rhetorical construction of economic value—what Thrift has called "the romance" of the new economy—and people's cultural evaluations of their labor, their companies, and the impact of economic forces in their lives. What is clear from the way people told stories about their experience in the crash is that the same elements of choice, freedom, and personal expression through career pathways in the new economy also led to self-blame and doubt during the crash. While people used technical, creative, and financial frameworks for evaluating their work in Silicon Alley, few could predict which particular new technologies or companies would succeed. The blustery rhetoric that the Internet would spark media, financial, and creative revolutions effectively silenced concerns about other potential negative outcomes or the ramifications of rapid growth in the sector. At one level of analysis, what happened at the end of the boom was a clash of valuations—a conflict among the multiple ways in which people valued the work, the products, the companies, and the potential of Internet-related companies. While multiple values got incorporated into constructing the market, financial values ultimately held the most sway and power within the field.

While it was functioning, this system of multiple valuations meant that people could interpret their work, their company's mission, and their characterization of the industry in order to fit the values that they held most dear. The historic cultural and economic changes that gave rise to the entrepreneurial spirit of the dot-com boom meant that decisions did not have to be tied tightly to financial justifications, even if they were mobilized in the service of capital. The autonomy that Ross wrote about in *No-Collar*, in which young workers could have meaningful jobs, coexisted with the needs and requirements of financial capital. Thousands of people were inspired to join small, start-up firms by rhetoric that encouraged entrepreneurial behavior. They joined for different reasons and had different ways of justifying the risks to themselves. These justifications collectively functioned as an emergent social structure to support risk taking. These justifications for risk were were flexible enough for the multiple ways in which people wanted to interpret it—from the cyberslackers to the e-business suits; venture capitalists and venture labor alike—all could embrace *something* in Silicon Alley. The industry supported these ways people evaluated worth and risk, even when these two contradicted each other. In a new industry, with fewer bureaucratic and organizational structures, uncertainty was organized by a powerful mechanism of risk taking that was culturally individuated and flexible enough to adapt to changing conditions within the economic terrain.

The risk within the Internet industry was not inherent in new endeavors, nor was it a natural outcome of new technological advances. Risk wasn't simply the product of the economic changes implied by the term "new economy." From job advertisements to popular press articles, there were clear ways in which employees were encouraged—even required—to partake in entrepreneurial attitudes as part of their jobs. Risk seemed natural, unavoidable, and not such a big deal. The corresponding rise in entrepreneurship seemed natural, too, and was one of the mechanisms for organizing the uncertainty of a new industry. Risk taking across the high-tech sector was perceived as being a requirement for working in the industry, not just a lifestyle option.

The dot-com crash also shows how people revert to a lens of individual experience to explain changes to economic structures. For people in Silicon Alley, their aspirations were quite different from the notions of the "American Dream" for a previous generation—they valued creative and economic freedom, independence, and making an impact. They framed economic success and failure using very individual terms when talking about their careers, constructing narratives of individual choice and responsibility to represent their particular economic consequences. Even though observers laid the blame for the crash on speculative capitalism, people in Silicon Alley represented their economic situations as something *personal* and something over which they had control. Taken together, these narratives present the recurring theme that success in the new economy is based on taking chances and making lucky choices—as opposed to, say, hard work, determination, loyalty, or a multitude of other possible narratives about economic success. People working in Silicon Alley may have overestimated or misrepresented their chances at success, however they defined it. They may have had incomplete information or misjudged their personal capacity to affect change within their companies. These valuations worked together to help create a situation in which economic growth seemed inevitable, but people blamed themselves for not being savvy enough when that turned out to be false. In this sense, I don't mean to argue that people are necessarily driven solely by logics of economic rationality, but rather that this notion of choice became a powerful way that active agents turned away from looking at other structural positions and possibilities for success and failure. Rather than placing blame on the economy or the stock market, people blamed themselves, suggesting on some level that they thought that they might be able to outwit the market.

Ultimately the financial valuation of the work that people were doing in Silicon Alley was different than the valuations they themselves placed on the jobs they were doing. Dot-commers in Silicon Alley grappled with a problem that generations of workers from every sector have had to face—financial capital is much more mobile than labor and better able to diversify across industries, countries, and businesses than people are able to diversify their human or their social capital.

Taking seriously the powerful cultural perceptions and narratives that workers used to frame their jobs leads to very different political implications. People in Silicon Alley actively welcomed and sought out risks, because risk is now tightly linked to what it means to be successful, creative, and in control of one's career. Creating better, more secure work for knowledge workers in general must seriously address this shift in the cultural landscape. A culture of risk survived the dot-com crash. Even though the Internet industry (as constructed during the first wave of dot-coms) failed, the entrepreneurial spirit that fostered it lived on. For high-tech firms and start-up Internet companies, skyrocketing stock prices during the late 1990s gave risk the shiny luster of potential wealth for employees, justifying in individual terms both the profits and losses that came with the stock market crash in 2000. Even though the IPO frenzy ended, the processes of venture labor continued—and continue—to function.

6 Conclusion: Lessons from a New Economy for a New Medium?

In December 2000, just a few months before the stock market crash that effectively ended the dot-com culture of Silicon Alley, Josh Harris, the founder of online video streamer Pseudo, told *60 Minutes* that his goal was to put the CBS television network out of business. At the time, less than 5 percent of American homes had broadband Internet access, and Harris's claim that web-generated and distributed video could one day threaten the dominance of an old media network sounded ludicrous. Harris's bravado came to symbolize the arrogance and brashness of Internet entrepreneurs. Journalists reported on Pseudo's extravagant loft parties full of downtown characters, and at one point people lived in the company as part of a performance art installation. The larger-than-life public persona of Harris and the image of Pseudo symbolized both the creative potential of Silicon Alley, as well as the excesses. A year later, it was Pseudo, not CBS, that no longer existed. Even though many Silicon Alley companies went out of business, the challenge that Harris—and new media more generally—presented to mass media now seems clear given the impact that YouTube, file sharing, and other online streaming content have already had on the on our media landscape.

People working in Silicon Alley believed strongly in the transformative power of the Internet, even if it was difficult to predict which particular companies would succeed. As one analyst put it when Pseudo announced its closure, "Only when we see one of these companies succeed will we be able to say, 'Oh, that's what Pseudo was missing.'"[1] Of course, now we know that one such company is YouTube, which was sold to Google for $1.65 billion in 2006. Pseudo may not have put CBS out of business, but just as Josh Harris predicted in 2000 online streaming video content has already drastically and irrevocably changed the economics of broadcast media. Even though Pseudo failed as a company, the idea that the company pioneered—video via the Internet—is now very much a central part of our media landscape. Silicon Alley was filled with companies like Pseudo that created tools for new media and concepts with significant financial, economic, and cultural potential. At the time, Steve Vondehaar, an analyst with The Yankee Group said, "Pseudo just ran out of time . . . they'd been around a long time and just couldn't wait around another three years for the market to develop. . . . I guarantee there's a new set of players out there preparing to be multimedia Webcasters, flush with fresh war chests and ready to throw money into the streaming media consumer space."[2]

And indeed, as YouTube's financial and popular success shows, there was. The "new set of players" also included more people who were, post-dot-com crash, willing to work for start-up Internet companies—a resource that is as important for success as money and time and timing. The same could be said, however, for almost any of the business and content areas in which Silicon Alley companies specialized. Many of today's successful Internet-related companies have predecessors in the first social networking sites, blogs, online advertisers, and magazines developed in Silicon Alley.

Pseudo's story reflects several aspects of Silicon Alley's legacy. The kinds of companies that remain, the impact on media, and the way in which entrepreneurial risk has changed are all encapsulated in this tale. In many ways, Pseudo symbolizes the first wave of entrepreneurs and venture labor for whom the crash did not represent the end of the Internet industry, but the beginning of a continuing story of new ventures and a continuation of the shift toward increased entrepreneurial work. The continued success of ideas that were first tested in New York connects the first wave of Internet industry venture labor to current and emerging web successes.

Looking back at the creation of Silicon Alley helps us bridge old and new economies, old and new media, and old and new ways of thinking about work. How people worked as venture labor in Silicon Alley helps us understand the ways in which the labor of producing culture has changed and will continue to change. In this light, these changes in work and in media are at once both fundamentally transformative and continuous in a historical trajectory of postindustrial, digital era developments. The business history of Silicon Alley is relatively overlooked, and yet because of New York's importance in Internet publishing and online advertising, this history is also critical for understanding how the Internet evolved and how it is still evolving. Many of New York's content-oriented Internet companies were founded with an explicit mission to create new forms of culture, forms that opposed those of corporate media or challenged corporate workplace cultures and hierarchical organizational principles. And yet what I have tried to show in this book is how people straddled multiple culturally framed calculations of value, such as financial worth and creative worth, in Silicon Alley, and how these different valuations functioned as a mechanism for dealing with economic uncertainty. Financial worth did not simply trump creative worth. Nor did these independent cultural rebels merely sell out for the inescapable allure of potential financial success, exchanging "bootstrapped" or self-financed business models for an association with the burgeoning dot-com financial phenomenon. Silicon Alley's legacy of media and of work can provide lessons for the future.

The Media Legacy of Silicon Alley

Silicon Alley left a legacy of downloadable songs, video streaming, Internet radio, social media, pop-up ads, and heartfelt personal blogs, all of which continue to reverberate in online media content. Videos are now commonly embedded in newspaper Web sites. Online tools make it easy for many people to start a blog or create a simple webpage. Social networking sites have changed how we connect to one another and use media in our daily lives. If anything, users have come to expect much more sophisticated and customized content. The revolution may not have been won by those who started it—namely, small, upstart independent online media companies like those in Silicon Alley—but the vision that they had changed the landscape of the Internet then and the expectations that we have as users of new media now.

The first wave of Silicon Alley firms launched several successes. Google bought DoubleClick, Silicon Alley's online advertising services company, for $3.1 billion in 2008. The women's news and information portal iVillage was sold to NBC Universal in 2006 for $600 million. Other content-oriented companies that were financially successful include Daily Candy and Nerve, both of which are still being published and are thriving. In 2003 Robert Pittman, a former executive of MTV and AOL, bought a majority stake in New York–based Daily Candy—an email newsletter on fashion, dining, shopping, and travel—for $3.5 million. In 2008 he sold his stake to cable company Comcast for $125 million.[3] Revenues of Nerve, Silicon Alley's highbrow adult content web site, increased more than fivefold from 2001 to 2006.[4] The former publisher of *Silicon Alley News* created Weblogs, Inc., a blog publishing company that he in turn sold to AOL for $25 million in 2005.[5] Silicon Alley has also produced several successful social media and Web 2.0 companies. Meetup.com rose to notoriety as the platform for linking Howard Dean supporters and activists during the 2004 presidential campaign and garnered venture capitalist investments from first-wave stalwarts Esther Dyson, Draper Fisher Jurvetson, Union Square Ventures, and eBay. Silicon Alley firms did not simply disappear with the dot-com crash, even if the industry was changed radically by it.

Part of what makes Silicon Alley an interesting case in which to see the changes in postindustrial work is that a group of young idealists were drawn to the industry by a mission to change media culture, where they faced the intersection of emerging media technology and economic trends of cultural production. Silicon Alley, for a while, transformed media work into new media work. As Peter Manuel showed in *Cassette Culture*, the introduction of new technologies can have a tremendous unforeseen impact on the structure of careers and industries within media, not just on the production and distribution of media content.[6] Silicon Alley expanded the definitions of high-tech work to include graphic design, writing, and art. A growing, cutting-edge industry and the dot-com boom, in New York at least, increased the demand for media-centered jobs. Digital media technologies may have some "disruptive" effects. But unlike Josh Harris's prediction that online video would put CBS out of business, new media has not completely dissolved the power of the old. If anything, Silicon Alley start-ups and other Internet industry companies revitalized and expanded the demand for music, stories, and video, creating new outlets and ways of viewing, interacting with, and engaging with media. The new media environments designed by the first wave of Internet pioneers created media and cultural landscapes that fostered laboratories of innovative ideas, in which relatively low-cost dissemination could reach unprecedentedly large audiences. The technological affordances of the Internet expands what Howard Becker has called the "art world"[7] of people engaged in media production, making visible many more ideas and concepts to a broader audience much earlier in the process. In such a media landscape a single writer's blog could lead to expanded audiences for other, related cultural products, as has happened for blog/book/movie adaptations such as Julie Powell's 2002–2003 blog "The Julie/Julia Project," which she adapted for a book that which was later made into the movie *Julie and Julia*.

The advances in social and economic organization that mark the digital age have changed and will continue to change our mediated lives radically and fundamentally. But while technology is certainly implicated in these changes, it is not the sole force behind them and the overarching social, political, and economic postindustrial shifts are not synonymous with technological ones. New media represents not only new technologies but new aspects of media culture.

We are only now beginning to understand the implications of those changes. Kevin Kelly has argued that once digital texts want to "weave themselves together," creating new links and convergences, with a power and force driven and desired by technology.[8] However, the distinction between production and distribution in several media is quite sharp. Record labels, for example, primarily serve as a conduit between independent cultural producers—musicians themselves—and an audience. This has always been true to a greater or lesser extent in cultural industries. Richard Caves has shown that contracts in cultural industries shift risks away from companies onto cultural producers themselves, a process that Paul Hirsch described as "filtering new products and ideas from 'creative' personnel . . . as they flow to the managerial, institutional, and societal levels of organization."[9] The Internet may have brought us self-made YouTube videos, but what remains after the dot-com euphoria is a tension between individual producers of creative ideas and the machine of cultural industries that produces, promotes, and distributes those creative ideas on a mass level. There may now be more reality television stars and more talent originating online crossing into traditional media, but their visibility and legitimation as stars still come from mass media attention.

Even if traditional media gatekeepers still control access to wide mass media distribution, individual media workers can test ideas online, often with little of the formal gatekeeping of traditional media organizations and institutions. This dynamic makes cultural production more democratic, bringing more and diverse voices into the reach of a potentially broader audience. At the same time, economic pressures stemming from digital media have decreased the diversity of production in almost every traditional medium. As a result of the new media revolution to move content online launched by the likes of Pseudo's Josh Harris, fewer print newspapers are published, record labels are introducing fewer new acts, and movie studios are more conservative in approving original projects. While the testing grounds for media talent may have moved online, the expansion of the Internet as a medium for content has had negative ramifications for the diversity of mass media. Rather than becoming more democratic, the means of mass media production and distribution appear to be even more tightly controlled. As Mark Deuze has argued, "disruptive" technologies like the Internet amplify and accelerate the sense that people have that we are living under constant change, in part because we are continually "living in and through media."[10]

The processes of corporate control of Silicon Alley described in this research challenge commonly held critical stances of media control. Self-proclaimed "cyberslacker" Jaime Levy could pronounce that the money and the business-oriented culture that took over Silicon Alley were "the Death of the Web as we knew it!"[11] even as she herself sought funding for the creative projects at her company Electronic Hollywood. I argue here that the cooptation of the values of independence of the early Internet pioneers began well before the first initial public offering in Silicon Alley (and well before the frenzy of the stock market bubble). When pressures emerged from investors for culture's profitability and for the rapid growth in the rate of those profits, creative risk taking was no longer a sufficient justification of value for any small firm within Silicon Alley, regardless of its level or type of funding. Financial accountability and business values were adopted as requirements for being considered legitimate within the field by content creators, even those not receiving venture capital or corporate funding. These shifting values in Silicon Alley were the result of the tight relationships early on between cultural and financial valuation. This process can be seen as an outcome of the social networks that embedded independence in Silicon Alley within a much larger ecology of corporate values, and not merely the result of changing ownership structures. New frameworks are needed to think through the distinction of production and valorization of media content in the market, with a particular focus on the labor that produces that content and how they frame their labor experience. The experience of labor producing content for the early commercial Internet does not fall neatly into Tiziana Terranova's concept of "free labor," nor is that labor wholly exploited, coopted, or an "industrialized bohemia," to paraphrase Andrew Ross.[12] Rather, the history of Silicon Alley means that we need to revisit our theories of the political economic complexities of cultural production.

   This book has examined the early history of the commercial World Wide Web through the entrepreneurial practices of people who shaped it. These young people created the Web as we now know it in that they were the ones who invented the first banner ads, the first online videos, first prototype blogs, and the first online stores. The early web had little in common with what we experience as users today, and the web is still evolving. However, the experiences of people working in Silicon Alley in the 1990s offer many important lessons for understanding technologies of both the media and the economy. The young pioneers of Silicon Alley tried to directly challenge the power of mass media, and they have already made their indelible mark on the way we buy and read books, listen to music, and watch videos. In short, what we see from the early years of Silicon Alley is how our mediated lives will be changed through the continued work of venture labor actively engaged in making new ways of interacting with media content. The lessons from the ways in which work emerged online in the dot-com era—the first generation of online digital laborers—help us understand the motivations and challenges of what might be called the social media era. They also show how the structures that once provided security within the workplace have been replaced by pressures for increased risk taking. Rather than utopian and dystopian visions of risk and security, the history of Silicon Alley shows the complexities of choices facing contemporary labor.

Lessons for the Workplace

Venture labor gives us a conceptual hook to understand changes to the nature of work and to understanding evolutions in new media. Venture labor, work that entails bearing entrepreneurial risk, is an explicit continuation and intensification of trends that began with deindustrialization. While "venturing" is not new—just think of the many who staked potential fortunes on the gold rush—compensating workers with stock options for part or all of their salary is relatively new, at least below upper-management level. Work in cultural industries has always been riskier for the people working in them. But what I have tried to present in this book are the financial implications of the shift of the burden of risk onto employees, the implications for media making in the digital era, and the potential of the application of the concept of venture labor to other industries. This suggests that venture labor is in part a response to changes in workplace security, the increased individualization of risk through health and retirement benefits, and structural changes across the economy.

One of those lessons is the changing relationship people have to risk. Several theorists including Ulrich Beck and Nicholas Luhmann have thought about the increase in risk in contemporary society.[13] But risk in these accounts is usually framed as a social bad—there's a veritable cottage industry in talking about the precarity of labor. Far from dismissing these claims, based on how people in Silicon Alley represented risk in their own lives, I argue that another level of complexity needs to be added to the existing theories of risk. People aren't simply suffering from false consciousness about the economic risks that they take. In fact, in the dot-com era entrepreneurship was something celebrated as cool and desirable. People willingly shoulder risks and this has—I fear—political ramifications that we have yet to fully think through. Whereas Beck and Luhman see the increasing risk within jobs as something to be avoided, many people whom I spoke to either actively welcomed risk or shrugged it off as a potential side effect of a much more desirable outcome, choosing not to frame risk in their depictions of it as something bad for them, even after the risks they took cost them their jobs, their companies, and sometimes their careers.

This is perhaps the single most important lesson that we can learn from Silicon Alley: the extent to which people's notions of job security have been radically transformed as more and more people *willingly* accept or actively welcome risk. People in Silicon Alley spoke of their careers as a series of risks—and blamed themselves when they did not take enough risks, or when the risks they took turned out to be bad ones. Risk became a conceptual lens through which people could view and represent their varied career choices. Even people using risk to describe creative chances they took or projects they developed played into an industry-wide social machinery that substituted talk of security for talk of calculation. In doing so, these cultural framings became the way in which risk was naturalized and managed.

The different frameworks that people used in Silicon Alley to manage risk were collectively part of an entrepreneurial environment—venture labor did the work of venture capital, except they took those risks *within* their own jobs. To those who say that young dot-commers had nothing to lose, that is wrong—people lost their jobs and their social capital. Their human capital investments in dot-com skills and training became practically worthless with the dot-com crash. These losses were not necessarily borne by individual workers either. In a Marxist sense, there was a "reproduction" of venture labor, with domestic arrangements bearing the brunt of the uncertainty that venture laborers face. Venture labor relies on support from domestic situations—being young, single, and able to fall back on family resources when things crash. These domestic resources ready flexible workers for the rigors of uncertainty and function to "reproduce" venture labor for the service of capital.

But still, people felt themselves to blame when the downturn happened. They had made choices; therefore, they felt themselves—not the economy or their companies—to blame. They told narratives and represented their positing using the language of risk and calculation, *even* when the underlying values of how they framed that risk were not financial. That is, people using the cultural or actuarial approaches to risk also appropriated the language of finance to talk about their careers, their passions, and their work lives. To varying degrees, individuals approached uncertainty as a problem to be solved and positioned themselves at the cusp of creative and financial trends, demonstrating that they struggled to navigate vagaries of market forces, even in the boom of the industry. Individuals made sense of the uncertainty of the field by using the tools and language of risk along with their own mechanisms of calculating worth. Individuals experienced economic uncertainty as a responsibility that befell them alone. This led to a continual fear of falling behind technological trends, a constant need to engage in after-hours networking to maintain social ties, and entrepreneurial pressure to outwit the financial markets, even when the industry was at its peak.

People working in Silicon Alley did not necessarily feel that their efforts protected them wholly from the vagaries of market forces, but they did feel a sense of control through the mechanisms they used to deal with economic uncertainty. When uncertainty is filtered through these culturally informed individual frames, individuals' stratified positions within a social structure are reaffirmed, not challenged, because uncertainty is managed as individual risk, not collective exposure or socially structured change. It is in this way that the cultural individualization of economic uncertainty can serve to exacerbate existing inequalities among workers. As one respondent in chapter 3 said, he was able to get ahead *because* he was willing and able to absorb risks, not in spite of them. Success, it now seems to these young workers, is the result of risk taking.

This was the first generation of workers entering the workforce after what Jacob Hacker has called the "great risk shift." What is interesting is that people welcomed risk. Risk was framed as inevitable, embraced by many, and feared by few. The narratives of people working in Silicon Alley reveal that these workers didn't see risk as bad, even when it meant bad outcomes for them. This research has shown that individuals are quite adept at justifying and accepting increased uncertainty, giving it narrative structure that puts them at the center of control and choice even when their jobs, careers, and the value of their accumulated social and human capital are at stake. Given the tenacity of these individual frames for managing uncertainty *and* how they are inextricably linked to conceptions of personal worth, it will be exceptionally difficult for new, collective forms of governance to emerge to handle increased uncertainty. This contradicts the predictions of Ulrich Beck in *The Risk Society* who sees the potential for new social mobilizing emerging out of the increased riskiness of contemporary life. Especially within the economic realm of decision making, risk is deeply held as being within the purview of individual choice and responsibility. The dot-com era proved to be no different.

Applying Venture Labor

The concept of venture labor, as a model of employee entrepreneurship, can be applied in a wide range of settings other than dot-coms. Companies starting up in any sector need the work of people who are committed to their projects, the company, or career potential within the industry and who are willing to invest time into the new venture. Beyond start-ups there are many types of work that require employee entrepreneurial investments of time, labor, or social or human capital into the products and services of the company in exchange for uncertain future payoffs.

Other communication and creative industries provide parallels for the work experience in Silicon Alley. Several of these industries require multiple unpaid internships as a de facto requirement for entry into a paid position. Usually considered training opportunities, unpaid work on the part of potential employees as entry into a firm or field could be seen as a form of venture labor. Such internships are on the rise among college students and recent graduates.[14]

Communication, media, and cultural industries also have a long tradition of working on "spec" or speculative forays into the creation of specific products. This is a special category of venture labor in which people invest time with the intent to sell the products of their work in a cultural marketplace dominated by gatekeepers and intermediaries. Jason Toynbee termed this phenomenon in music "outworking, whereby musicians carry the burden of buying their own equipment, of writing and rehearsing, and then absorbing the risks of an uncertain . . . market until the unlikely event that they are recruited by a record company."[15] William Goldman famously wrote "nobody knows anything" in the film industry, a reflection of the frustration of a writer to get approval for the ideas she or he speculatively pitches to studio decision makers.[16] The "cultural industries" and media economics approach to work in media such as that of David Hesmondhalgh and Richard Caves emphasizes the risks that creative workers in these industries face.[17] Using the concept of venture labor to understand these risks highlights the entrepreneurial nature of the work that is required as well. This work is not limited to creative employees. Postproduction film editors, for example, require specialized training, and staying employable means keeping abreast of rapidly changing technology and making and maintaining contacts in the industry.[18] Maintaining future employability in creative industries often requires years of unpaid training, continual skills updating, and market forecasting on the part of prospective employees, in addition to any specific investments in a particular project or company.

The concept of venture labor helps frame both the conditions within a particular job and the context in which work now occurs. The narrative of individual investment into careers and jobs is now pervasive throughout the economy. Jared Bernstein has called this thinking "YOYO" economics, in which "you are on your own."[19] Such narratives are part of a cultural transformation of work and workers into even more highly commodified products and of the necessary components into specific "investments" with returns. For example, Vicki Smith and Esther Neuwirth show the historical construction of the "good temp" rose in this period, in which the market for an hourly flexible worker was constructed by the temporary help services industry and was normalized in an employment framework that is "insecure, remains impermanent across jobs and occupations, and places high degrees of risk on workers."[20] Similarly, getting an education has taken on more of the narrative of an investment in the future, and, as Anya Kamenetz in *Generation Debt* showed, thinking of students' educations as their investments has helped contribute to the recent skyrocketing of student loan debt. Now, in the United States the total amount of student debt exceeds the total amount of credit card debt as increasing education costs are sold to students as investments in potential future careers.[21]

One of the catastrophic outcomes of increased financialization, privatization, and neoliberal ideologies is that investment metaphors will continue in different settings and different markets. The dot-com dream promised workers the chance of becoming millionaires (or "thousandaires" as several respondents put it in our interviews) from investments of luck and labor time. After the dot-com boom and bust came the U.S. housing bubble. Housing market dreams, too, promised middle-class families stellar returns on their investments in homes, which a cultural machinery of home-flipping television shows, investment columnists, and news articles endlessly repeated as a source of potential riches. However, just as with venture labor, these investments couldn't be easily diversified across multiple homes or jobs, and the hedges or protections for these investments are not as strong or flexible as those for financial capital.

For example, social networks were a significant source of support for workers in Silicon Alley, and such networks tightly link together firms within an industry and region. However, the downside of such tight links was that they provided little support when the system began collapsing. They are also costly in real terms to the workers who acquire them, meaning that social networks require time that could be spent doing other things. The creative, cultural industries as engines of regional growth as described by urban theorists Richard Florida and Elizabeth Currid depend on after-hours work.[22] I show in chapter 4 that these are investments in companies and regions, and they cost the people who build them in terms of time and work. However, my research points to a much darker side of the highly linked urban economies of cultural production. There is clearly a great degree of *work* involved in building and maintaining these regional economies, and this work is disproportionately done after hours by people who have the time, ability, and social capital to navigate such events.

Networking among the employees of an industry forms a paradox of social capital creation in which individual resources benefit companies and industries. However, as the downturn in Silicon Alley showed, the creation of this social capital was an effective strategy for managing the risk incurred by and at individual companies, it did little to help buffer workers against the "systemic risk" of an industry downturn. Although several researchers have focused on workers' social networks as new forms of "labor market intermediaries," there has been little research that connects individual-level social networking with broader industry effects, as I have done in this study.[23] The data on events held in Silicon Alley shows that links among different sectors of Silicon Alley changed over the course of history. This suggests not only that workers were acting as agents of their own interests in making connections for new jobs and obtaining new information at these events, but that they also served a function in maintaining ties necessary for tightly linked regional economies. The interviews with people who were directly involved in promoting their companies confirms the pressure they felt to go out. But interview data from workers at lower levels in their companies also shows how these ties were built, maintained, and used, both by individuals looking for resources for their own careers and by their employers. Of course, for freelance or contract employees, this was even more the case because they were expected to bring outside knowledge from the rest of the industry into the companies in which they worked.

Consequences of Financialization and Privatization

From slashed benefits to contingent pay, work within the American economy became more, not less, unstable during the latest economic boom. The entrepreneurial spirit of the dot-com boom showed a positive side of this insecurity, promoting acquiescence and acceptance of an economic regime that benefits the few and holds individual responsibility for market failures. Although the fast-paced work environments of Silicon Alley are an extreme case, those who worked there were in many ways what Andrew Ross has called "avatars of uncertainty."[24] Given that the jobs being created in high-tech boomtowns across the country were heralded as the future of work and as being humane and even fun with an even greater deal of worker autonomy, the fact that these jobs were associated with a great deal of turbulence does not bode well for the American economy's ability to create stable jobs. During a booming economy, one of the fastest-growing sectors provided employment for a relatively narrow sliver of workers. And this workforce accepted, in varying ways, the uncertainty and capriciousness of their work, even as their sector was expanding. Their dreams and hopes were mobilized in order to attract people to work in a new field. They traded security for risk, safety for opportunity.

My fieldwork uncovered the cultural frames that workers used to calculate value in their jobs—and by extension to frame and manage the risks they perceived they faced. Their acceptance and management of risk served capitalist functions by supporting the growth of companies and industries. In effect, their individual-level, culturally informed explanations that my respondents offered for how people get ahead, why work is meaningful, and how to navigate a career in this field masked the social or collective forces at play within their jobs. This does not bode well for creating a sense of old-fashioned solidarity among workers facing increased uncertainty. To echo the words of one of my unemployed respondents after the crash, when a job is thought of as an investment—and not as a right, an obligation, or an earned position—losing a job can seem like "easy come, easy go." A sense of moral outrage or indignation at being part of an economic arrangement that devalues work and views, in Louis Uchitelle's terms, workers as "disposable" is missing when people use these investment frames to understand their choices.[25]

People invested their passion, their time, and their efforts only to feel betrayed when they realized the great business revolution they thought they were creating was merely business as usual. However, the Marxist concepts of self-exploitation and false consciousness do not adequately explain the dynamics of the choices people made. Workers could not necessarily access the larger structural position they were in. That is, they could not necessarily perceive the ways in which the economic environment had shifted, as my respondents taught me when I asked them for their views on job security. From within, their choices made sense to them—both in terms of a rational economic sense and a social cultural sense. They became part of an entrepreneurial moment to which they were called and found recognition in a way of life, a style of work, and a mode of being that made perfect sense to them and within the social structures they perceived. Still they lacked the ability to access why or how they fit into these social structures—how their individual choices played into and were shaped by larger social dynamics.

Venture labor stood in for venture capital—both in the real sense of resources needed to build companies and in terms of seemingly being an accessible option for wealth accumulation of middle-class knowledge workers. This plays into the political right's rhetoric of the "ownership society" that employs the middle-class dream of profiting from ownership. Whether from one's labor or one's home, the false dream of the ownership society promises that generating profits in a capitalist society is as easy as cashing a paycheck or paying the mortgage. So-called ownership society ownership comes without power, just as the stock option holders of Silicon Alley exercised little power over their companies' choices. Similarly, the U.S. housing market crisis exposed the culpability of middle-class investments in a global financial marketplace.

Financial capital does not have to be directly involved for financial values to become incorporated. Financial capital came to Silicon Alley relatively late in the formation of the industry. However, the myth of Silicon Alley's early independence does not hold up to historical scrutiny. The tight and early linking of corporate media interests to New York's Internet industry lays to rest the nostalgic view of a wholly alternative or independent early milieu of the production of culture for the Internet in Silicon Alley. The social process by which Internet production moved from high-cultural, elite, artistic cultural production to mass distribution did not entail, the "industrialization of bohemia." Rather, the process is better described as what Pierre Bourdieu has called a struggle between those who dominate a cultural field politically and economically and those who can articulate more autonomy within a field, creating "art for art's sake."[26] The struggles in the evolution of the development of content for the web to balance cultural innovation with mass appeal and accessibility echo the tensions between high and low culture in other cultural industries.

Nor is it as simple as blaming financial capital for the creation and collapse of the dot-com bubble. Individuals had some control over their careers and lives. Young workers in Silicon Alley exercised more ability relative to much of the labor market to make choices, take chances, and have responsibility, and theirs were very good jobs, relatively speaking. They weren't duped into taking them; these jobs stacked up nicely compared to the other alternatives they saw in the economy. However, the dot-com boom serves as what Clifford Geertz called a "paradigmatic human event," one that tells us less about a particular universal truth, but more about the process that could and does happen again in different kinds of settings.[27] The story of the dot-com bubble is less about how one particular industry generated an entrepreneurial moment, but rather how such moments get shaped by and reflected in the narratives we tell ourselves and others. Booms draw people in and keep them there, and in doing so, function to disperse risks across a greater number of people. The lesson from the dot-com boom is not that people are good or bad at calculating these risks, but that different sets of parameters for gauging risks can mean that individual choices serve a function: to diminish the appearance of risk. In the case of the dot-com boom, that function was to create a sense of opportunity, expansion, and security amid a labor market that was fundamentally destabilized. The next boom, and the next inevitable bust, will work similarly, with equal or greater devastation.

Going Forward

Those who produce culture and technology continually face the pressure of needing to figure out "the next big thing." Scholars who wish to understand how media and cultural products look the way they do will need to rely on inquiry into the structures of the production. My call to other communication and media scholars is to take up research on these social, organizational, and institutional structures to help explain media landscapes. The entrepreneurial pressures upon workers to figure out market trends, accept uncertain work, and tie their pay to the financial success of their products, play, I suspect, an even greater role in the organization of cultural industries—and the way that those media products look—than researchers now recognize. This book focused on one particular media industry: that of content for the early commercial Internet, which was an extreme case in terms of the volatility experienced within it. However, the history of the dot-com boom suggests that people interested politically in creating better, more stable jobs will need first to tackle the enormous task of figuring out new, collective ways to organize uncertainty within economic production.

Emile Durkheim in his classic work of sociology, *Suicide*, found that the apparently most personal of decisions are still the products of a wide array of social influences.[28] Work and career decisions are no different—economic decisions are not made in isolation, and great entrepreneurial rushes do not occur outside the collective forces that shape social life. In general, the language of economics provides very little scope for challenging the apparent naturalness of individualism. This book has shown how the social and cultural practice of framing, understanding, and calculating risk was necessary for managing the uncertainty of a new industry. The story of Silicon Alley, a small but important and highly visible regional economic sector, reflects how economic phenomena like risk are deeply embedded within cultural practices and how industries depend on the communicative practices of the people working in the field.

Beck, among many others, theorized the increasing "privatization of risk" in contemporary economic life. As an empirical study of how individuals experience and frame economic uncertainty in one specific field, this research is able to contradict some of the hypotheses of the *Risk Society* theories. Individuals understand, frame, and attempt to calculate economic uncertainty using the cultural tools available to them. For some in Silicon Alley, this was through mechanisms of understanding their creative contributions to a company or project. To others, it was through figuring out how to make themselves continually employable by minimizing their exposure to "risky" projects or companies. And for a visible few, risk was sought after as the only chance for garnering success in a turbulent labor market. These culturally informed stances—these different mechanisms of calculating worth—challenge extant theories of the privatization of economic risk in the workplace. These justifications of value evidence a process of structuration—of agents being shaped by and agents shaping the economic phenomena around them, rather than their wholesale acceptance or rejection of the increasing uncertainty inherent in modern employment relations. This is what Anthony Giddens means by structuration—that our microlevel actions are both constrained by and constitute the larger patterns of social behavior around us. Venture labor in the new economy was shaped by social structures that allowed the individual framing of risk to function in the service of capitalism. As Karel Williams has written, "The new economy matters because it was, and is, business as usual, acting out changes and continuities which are part of our future as much as our past and which have as much to do with finance and politics as with technology."[29]

We are living in a moment of extreme excitement about technological change. In many ways, the rise of Web 2.0 shows that Silicon Alley's vision of the Internet has, in essence, won out over other models of what the

Internet could become. Today's social media is even more driven by an ethic of self-publishing, by social connection, and by what my colleague Hanson Hosein calls a "storyteller uprising."[30] There will no doubt be a Web 3.0 or other such reinvention, expansion, or innovation in media and communication technologies. The trick for future media and business revolutions will be to find ways to support venture labor, so that innovative and creative jobs can also be stable and good jobs.

Notes

1 The Social Risks of the Dot-Com Era

1. The names of all respondents in the book have been changed to protect their confidentiality.
2. Geertz, *The Interpretation of Cultures*.
3. Schatzman and Straus, *Field Research Strategies for a Natural Sociology*, 53.
4. Anonymous, *Silicon Alley Reporter*.
5. Beck, *Risk Society*.
6. Van Slambrouck, "America's Growing Affinity for Risk."
7. Douglas and Wildavsky, *Risk and Culture*.
8. Knight, *Risk, Uncertainty, and Profit*.
9. Pham and Stoughton, "Dot-Com Firms Foster New Corporate Culture," A26.
10. Ibid.
11. Dyson, "So Many Ideas, So Few Companies," 16.
12. Durkheim, *Suicide*, 309.
13. Block, *The Vampire State*, 8.
14. Uchitelle, *The Disposable American*.
15. Hacker, *The Great Risk Shift*, 244.
16. Brenner, *The Boom and the Bubble*; Kripner, "The Financialization of the American Economy."
17. Beunza and Stark, "Tools of the Trade"; Bruegger and Cetina, "Global Microstructures."
18. Thrift, "It's the Romance Not the Finance That Makes the Business Worth Pursuing."
19. White, *Markets from Networks*, 299.
20. See Beck, *Risk Society*.
21. Castells, *The Rise of the Network Society*.
22. Smith, *Crossing the Great Divide*, 7.
23. Smith, "New Forms of Work Organization," 332.
24. Kanter, "Nice Work If You Can Get It."
25. Beck, *Risk Society*.

26. Hacker, *The Great Risk Shift*.
27. Pham and Stoughton, "Dot-Com Firms Foster New Corporate Culture," A26.
28. Child and McGrath, "Organizations Unfettered," 1145.
29. Frank, *One Market under God*.
30. Hacker, *The Great Risk Shift*, 67.
31. Ibid., 68.
32. Ibid., 40.
33. Van Loon, *Risk and Technological Culture*, 35.
34. Thrift, "'It's the Romance Not the Finance That Makes the Business Worth Pursuing.'"
35. Ross, *No-Collar*.
36. Turner, *From Counterculture to Cyberculture*.
37. Amoore, "Risk, Reward and Discipline at Work."
38. Hodgson, *Economics and Utopia*, 69–71.
39. Ibid., 181.
40. Ibid., 203.
41. Smith, *The Right Talk*, 65.
42. Van Loon, *Risk and Technological Culture*, 40.
43. Uchitelle, *The Disposable American*.
44. Hacker, *The Great Risk Shift*, 78.
45. Cantillion, *Essay on the Nature of Commerce in General*, I.VII.3.
46. Hacker, *The Great Risk Shift*, 75.
47. Castells, *The Rise of the Network Society*, 282
48. Sennett, *The Culture of the New Capitalism*.
49. Pink, *Free Agent Nation*, 10.
50. Peters, "The Brand Called You."
51. Mandel, *The High-Risk Society*, 105.
52. Peters, "What We Will Do for Work," 42.
53. Quoted in Massanari, "Dot-Coms and Cybercultural Studies," 421.
54. Neff, Wissinger, and Zukin, "Entrepreneurial Labor among Cultural Producers."
55. Douglas and Wildavsky, *Risk and Culture*.
56. Conversation with Mike Blain, January 2000. See also Blain and Wilson "Organizing in the New Economy."
57. Douglas, "Risk as a Forensic Resource," 3.
58. Ibid., 7.
59. Kait and Weiss, *Digital Hustlers*.
60. NYNMA and Coopers & Lybrand, "2nd New York New Media Industry Survey," 26.
61. Batt et al., *Net Working*.
62. NYNMA and Coopers & Lybrand, "2nd New York New Media Industry Survey"; NYNMA and Price Waterhouse Coopers, "3rd New York New Media Industry Survey."
63. Kalleberg, Reskin, and Hudson, "Bad Jobs in America."

64. Batt et al., *Net Working.*

65. Kunda, Barley, and Evans, "Why Do Contractors Contract?"; Kalleberg, Reskin, and Hudson, "Bad Jobs in America."

66. Saxenian, *Regional Advantage.*

67. Edgecliffe-Johnson, "From Dotcom to *Garçon.*"

68. Lyon, "Participants' Use of Cultural Knowledge as Cultural Capital in a Dot-Com Start-Up Organization."

69. Saxenian, *Regional Advantage*; Florida, *The Rise of the Creative Class.*

70. Castells, *The Rise of the Network Society.*

71. Girard and Stark, "Distributing Intelligence and Organizing Diversity in New Media Projects."

72. Rose, *Inventing Ourselves.*

73. Neff, Wissinger, and Zukin, "Entrepreneurial Labor among Cultural Producers."

74. McRobbie, "From Clubs to Companies," 518.

75. Lash and Urry, *Economies of Signs and Space*, 16.

76. Flew, "Creativity, Cultural Studies, and Services Industries," 187.

77. Ibid.

78. Ibid., 186.

79. Pew Internet and American Life, http://www.pewinternet.org.

80. Turner, *From Counterculture to Cyberculture.*

81. Howard, "Network Ethnography and the Hypermedia Organization."

82. Miles and Huberman, *Qualitative Data Analysis*, 102–105.

83. Stark, *The Sense of Dissonance*, 196.

84. Smith, *Crossing the Great Divide.*

2   The Origins and Rise of Venture Labor

1. Hamm, "Finding Your Way."

2. Quoted in Brown, "The '100 Best Companies.'"

3. Quoted in Levering and Moskowitz, *The 100 Best Companies to Work for in America*, 193.

4. Ibid.

5. Cappelli, *The New Deal at Work*, 73. To be fair, IBM began a program of downsizing their workforce before 1994, but these were through attrition and early-retirement incentives rather than layoffs.

6. Uchitelle, *The Disposable American*, 145.

7. IBM, "IBM Highlights."

8. Samuelson, "R.I.P."

9. Whyte, *The Organization Man.*

10. Drucker, *Innovation and Entrepreneurship*, 256.

11. Limerick, "Of Forty-Niners, Oilmen and the Dot-Com Boom."

12. Mandel, "The Digital Juggernaut," 22.

13. Mandel, "The Job Market Isn't Really in the Dumps," 42.

14. Mandel, "You've Got the New Economy All Wrong, Lou."

15. Reich, "Casualties of the Inflation War."

16. Cappelli, *The New Deal at Work*, 1.

17. Ibid., 2–3.

18. Uchitelle, *The Disposable American*.

19. Hacker, *The Great Risk Shift*, 45.

20. Pearlstein, "New Economy Gives Work a Hard Edge."

21. See, among others, Black and Lynch, "What's Driving the New Economy?"; Gordon, "Hi- Tech Innovation and Productivity Growth"; Stiroh, "Is There a New Economy?"

22. Mandel, "The Triumph of the New Economy."

23. Pearlstein, "New Economy Gives Work a Hard Edge."

24. Uchitelle, *The Disposable American*, ix.

25. Hacker, *The Great Risk Shift*, 38–39.

26. Houck and Kiewe, eds., *Actor, Ideologue, Politician*, 177.

37. Kiewe and Houck, *A Shining City on a Hill*, 217.

28. Ibid.

29. Houck and Kiewe, *Actor, Ideologue, Politician*, 270.

30. Ibid.

31. Ibid., 274.

32. Political scientist Mark Smith found that Republican Party platforms after 1973 were more likely to emphasize the economy than those before then. Smith, *The Right Talk*, 134.

33. Democratic Party Platform of 1992: A New Covenant with the American People, July 13, 1992. Available online from The American Presidency Project, http://www.presidency.ucsb.edu/ws/index.php?pid=29610 (accessed June 2, 2009).

34. Governor Bill Clinton, Speech on the Economy, Wharton School of Business, University of Pennsylvania, Philadelphia, April 16, 1992, Available online from Campaign Trails, http://campaigntrails.org/catalogs/1992-Clinton-FTP/wharto.txt (accessed September 9, 2011).

35. Ibid.

36. Ibid.

37. Mosco, *The Digital Sublime*.

38. Ibid., 13.

39. Frank and Cook, *The Winner-Take-All Society*.

40. Quoted in Pearlstein, "New Economy Gives Work a Hard Edge."

41. Pearlstein, "Reshaped Economy Exacts Tough Toll."

42. Cappelli, *The New Deal at Work*, 65, 67.

43. See the emerging literature on how economists and analysts "perform" economic markets, e.g., Mackenzie, Muniesa, and Siu, eds., *Do Economists Make Markets?*

44. Thrift, "'It's the Romance Not the Finance That Makes the Business Worth Pursuing.'"

45. De Cock, Fitchett, and Volkmann, "Constructing the New Economy," 48.

46. Williams, "Business as Usual," 399.

47. Clark, Thrift, and Tickell, "Performing Finance."

48. Thrift, "'It's the Romance Not the Finance that Makes the Business Worth Pursuing,'" 414.

49. Williams, "Business as Usual," 410.

50. Aglietta and Breton, "Financial Systems, Corporate Control and Capital Accumulation."

51. Glassman and Hassett, *Dow 36,000*; Elias, *Dow 40,000*; Kadlec, *Dow 100,000*.

52. Anonymous, "The Map."

53. That billboard came down in March 2002. Lee, "A Once-Evocative Name Falls Victim to the Bursting of the High-Tech Bubble," 14:4. It is interesting to note that this article was written as part of the weekly "Neighborhood Report" section of the Sunday *New York Times*. The Neighborhood heading is "Silicon Alley," not the Flatiron District or Chelsea, the more generally accepted names for the area.

54. Benner, *Work in the New Economy*; Saxenian, *Regional Advantage*; Florida, *The Rise of the Creative Class*; O'Riain, "The Flexible Developmental State."

55. Chervokas, "Design Shops Know."

56. Indergaard, *Silicon Alley*.

57. Chervokas and Watson, untitled editorial.

58. Chervokas, "Online Arts Are More Important to Silicon Alley than Venture Capital."

59. NYNMA and Coopers & Lybrand, "New York New Media Industry Survey."

60. Quoted in Deutsch, "On Electronic Highway, Manhattan Is a Destination," 9:1.

61. Chervokas and Watson, "The Year the Alley Grew Up."

62. Jason Calcanis, editor of *Silicon Alley Reporter*, quoted in Grigoriadis, "Silicon Alley 10003."

63. For more on the politics and culture of print 'zines, see Duncombe, *Notes from Underground*.

64. John Motavalli notes that the first magazine on the Internet was *The New Republic*, which was hosted on a gopher site. See Motavalli, *Bamboozled at the Revolution*, 50.

65. Ibid., 65–76. The Shannons claimed that *Urban Desires* was the first webzine.

66. Anonymous, "The New York Cyber Sixty," 51.

67. Watson, "Where Content Is King."

68. Anonymous, "Web Publishing Ideal Lives as Coalition Funds FEED."

69. Weil, "Can Consumer Companies Profit from Web Art Experiments?"

70. Ibid.

71. She, of course, included herself in this category. Quoted in "Interview with Jaime Levy," *Interface NYC*, January 1994, http://www.exhibitresearch.com/kevin/nyc/levy/index.html (accessed September 26, 2011).

72. Quoted in Chervokas, "New York New Media's Ground Zero."

73. Chervokas, "Design Shops Know."

74. Morrissey, "Primedia Scoops Up gURL.com From Delia's."

75. As of June 2003, two of the three founders, Esther Drill and Heather MacDonald, were still listed on the masthead for *gURL*. It is now published by Alloy Media and Marketing, http://www.gurl.com/about-gurl/ (accessed September 26, 2011).

76. See http://www.breakupgirl.net/friends/history.html (accessed September 26, 2011).

77. NYNMA and Coopers & Lybrand, "2nd New York New Media Industry Survey," 1997.

78. "@News: Digital City closes Total NY and Ada'web," *AtNewYork*, March 6, 1998. Its cofounder Benjamin Weil continues to be involved with digital arts projects as a curator for media arts at the San Francisco Museum of Modern Art and with the New York-based digital arts initiative, Eyebeam.

79. Chervokas, "In the Evolving Internet Industry, Web Designers Need Not Apply."

80. Ibid.

81. Anonymous, "Will Content Stay King of New York?"

82. Watson, "Where Content Is King."

83. Ibid.

84. Marisa Bowe, "Word Up! Content's Success Depends on the Right Model."

85. Watson, "Where Content Is King."

86. Robert Sikoryak, "Hacking It."

87. Anonymous, "Wired," 167.

88. Anonymous, "Net Work News," 171.

89. Anonymous, "The Breakfast Club," 222.

90. Anonymous, "The VCs Don't Get It and It's Our Own Fault."

91. See Centner, "Places of Privileged Consumption Practices."

92. Harmon, "Beyond Boosterism in the Alley," D1.

3   Being Venture Labor

1. Ross, *No-Collar*, 15.

2. Ibid., 20.

3. Boltanski and Thévenot, *On Justification*.

4. Ibid., 227.

5. Eliasoph and Lichterman, "Culture in Interaction."

6. Adkins, "The New Economy, Property and Personhood."

7. Ibid., 112.

8. Jason Calacanis, quoted in Indergaard, *Silicon Alley*, 82.

9. Candice Carpenter, quoted in ibid., 97.

10. Of course, many New York Internet companies have been "delisted" from stock exchanges after their stock prices dipped too low to remain eligible for trading.

11. Kushner, "New York Super Schmooze."

12. Batt et al., *Net Working*; Christopherson, "Project Work in Context"; Pratt, "New Media, the New Economy, and New Spaces"; Indergaard, *Silicon Alley*; Girard and

Stark, "Distributing Intelligence and Organizing Diversity in New Media Projects"; Grabher, "Ecologies of Creativity."
13. Frank and Cook, *The Winner-Take-All Society*.
14. Neff, Wissinger, and Zukin, "Entrepreneurial Labor among Cultural Producers."
15. Kanter, "Careers and the Wealth of Nations," 509.
16. Ross, *No-Collar*, 34.
17. Batt et al., *Net Working*.
18. Thrift, "'It's the Romance Not the Finance That Makes the Business Worth Pursuing.'"
19. Ibid.
20. Christopherson, "Project Work in Context."
21. Smith, *Crossing the Great Divide*.
22. Raj Jayadev, "Organizing Infotech Workers," presentation to the LaborTech conference, September 2002, New York, N.Y.

4   Why Networks Failed

1. Bernardo Joselevich, "Bernardo's List" email, March 7, 2001.
2. Bernardo Joselevich, "Bernardo's List" email, March 26, 2001.
3. McRobbie, "From Clubs to Companies."
4. Ross, *No-Collar*.
5. Pulitzer, "Razorfishsucks.com."
6. Pulitzer, "@The Scene," July 11, 1997.
7. Pulitzer, "@The Scene," March 20, 1998.
8. Gabriel, "Where Silicon Alley Artists Go to Download."
9. Pulitzer, "Three Day Rave."
10. Pulitzer, "Holy Holiday."
11. Currid, *The Warhol Economy*, 95.
12. Chet, "Like Exporting Baseball"; Centner, "Places of Privileged Consumption Practices."
13. Benkler, *The Wealth of Networks*, 3.
14. Ibid., 15.
15. Castells, *The Rise of the Network Society*.
16. Weinstein, "Make Networking a Career Habit."
17. Leonard, "Networking Still the Best Way to Find Work."
18. Gladwell, "Six Degrees of Lois Weisberg"; Gladwell, *The Tipping Point*.
19. See, for example, Batt et al., *Net Working*; Benner, *Work in the New Economy*.
20. Kadushin, *American Intellectual Elite*.
21. Grabher, "Collective Tinkering: Emerging Project Ecologies in New Media," 1919.
22. Batt et al., *Net Working*; Benner, *Work in the New Economy*; Christopherson, "Project Work in Context."
23. Anonymous, "1997's Silicon Alley Reporter 100," 5.

24. Cited in Tractenberg, "New York Grows as a Multimedia Hub as More Firms Set Up Projects There."

25. Joyce, "Alley Buzz."

26. For a fascinating account of WELL and the role it played in shaping digital culture, see Turner, *From Counterculture to Cyberculture*.

27. Reported in Pulitzer, "@The Scene: Why Silicon Alley Is a Real Community," *AtNewYork*, October 17, 1997.

28. The newsletter was known as *@NY* in its earlier years and often refers to itself in this shorthand. For consistency, I'll refer to the publication as *AtNewYork*.

29. From the masthead of the July 25, 1997, issue.

30. Chervokas, "Saying Goodbye to Atnewyork.com," *AtNewYork*, May 25, 2000.

31. Kadushin, *American Intellectual Elite*; Becker, *Art Worlds*.

32. Grigoriadis, "Silicon Alley 10003."

33. Anonymous, "The New York Cyber Sixty."

34. *AtNewYork*, September 20, 1996, and *AtNewYork*, November 1, 1996.

35. She also left her email address and a reminder about the fact that she hosted a discussion group on ECHO, the East Coast Hang Out, an early Internet provider and community site for many who worked in Silicon Alley. See "@The Scene," *AtNewYork*, March 21, 1997, http://web.archive.org/web/19980516041614/http://www.atnewyork.com/scene.htm (accessed May 20, 2003).

36. Zook, "The Web of Production," 418.

37. Pulitzer, *The Cyber Scene*, January 29, 1999.

38. Watson, "What the Industry Needs."

39. See also Granovetter, *Getting a Job*.

40. For more, see Neff, "Associating Independents."

41. Christopherson, "Project Work in Context," 2012.

42. Pulitzer, "@The Scene," November 14, 1997.

43. Currid, *The Warhol Economy*, 103.

44. Saxenian, *Regional Advantage*, 9.

45. See also Heydebrand, "Multimedia Networks, Globalization and Strategies of Innovation."

46. Saxenian, *Regional Advantage*, 33.

47. Shy, *The Economics of Network Industries*.

48. Piore and Sabel, *The Second Industrial Divide*.

49. Benner, *Work in the New Economy*; Batt et al., *Net Working*.

50. Coleman, "Social Capital in the Creation of Human Capital."

51. McRobbie, "From Clubs to Companies."

52. Ibid., 519.

53. Ronald Burt, *Structural Holes*.

54. Christopherson, "Project Work in Context," 2012.

55. Ibid.

56. Washington Alliance of Technology Workers, "The State of Seattle Area IT Employment and Training."

57. See Batt et al., *Net Working*; Washington Alliance of Technology Workers, "The State of Seattle Area IT Employment and Training."

58. McRobbie, "From Clubs to Companies," 522.

59. Hochschild, *The Managed Heart*.

60. Florida, *The Rise of the Creative Class*, 184.

61. Currid, *The Warhol Economy*, 4.

62. Ibid., 99.

63. Centner, "Places of Privileged Consumption Practices."

64. Batt et al., *Net Working*.

65. Ibid.

5 The Crash of Venture Labor

1. Slater, "From Calculation to Alienation," 244. For Marx, alienation happens when workers' products seem or appear independent of them. See Marx, "Estranged Labor," 69.

2. Marx and Engels, *The Communist Manifesto*, 38.

3. McRobbie, "From Clubs to Companies."

4. Motta, "Pace of Dot-Com Layoffs Slows in March."

5. Chervokas, "Billions Flowed into Silicon Alley, as Internet Industry Grew Up in 1999."

6. Ibid.

7. Chervokas and Watson, "Everything Old Is New Again."

8. Anonymous, "Is Gotham Ready?"

9. All classified ads come from *The Silicon Alley Reporter*, an online trade publication for New York's Internet industry. These ads are sampled from March 2000.

10. Anonymous, "Xceed Gains Former HBO Vice President."

11. Author's analysis of MBA course offerings from 1997 to 2002 at Columbia, Harvard, Stanford, MIT, NYU, Wharton, and the University of Chicago.

12. U.S. Department of Labor, *Occupational Outlook Handbook 2000–2001*.

13. Anonymous, "The Atnewyork.Com Index."

14. Anonymous, *Alley Cat News*, February 2001, 24.

15. Abreu, "The Beginning of the End of the End?"

16. Ibid.

17. Mandel, "You've Got the New Economy All Wrong, Lou."

18. Thrift, "'It's the Romance Not the Finance That Makes the Business Worth Pursuing,'" 412.

19. Kripner, "The Financialization of the American Economy."

20. Chet, "Like Exporting Baseball," 20.

21. Ibid., 21.

22. Slater, "From Calculation to Alienation," 248.
23. De Certeau, *The Practice of Everyday Life.*
24. Slater, "From Calculation to Alienation," 248. See also De Cock, Fitchett, and Volkmann, "Constructing the New Economy," 48.
25. Stark, *The Sense of Dissonance*, 11.

6 Conclusion

1. Alvear and Top, "Pseudo Bites the Dust."
2. Ibid.
3. St. John, "Alive and Well in Silicon Alley"; Kafka, "Comcast Buys DailyCandy for $125 Million."
4. St. John, "Alive and Well in Silicon Alley."
5. Ibid.
6. Manuel, *Cassette Culture.*
7. Becker, *Art Worlds.*
8. Kelly, "Scan This Book!; Kelly, *What Technology Wants.*
9. Caves, *Creative Industries*; Hirsch, "Processing Fads and Fashions," 642.
10. Deuze, *Media Work*, 233.
11. Grigoriadis, "Silicon Alley 10003."
12. Terranova, "Free Labor"; Ross, *No-Collar.*
13. Beck, *Risk Society*; Luhmann, *Risk.*
14. Gina Neff, "The Competitive Privilege of Working for Free," presentation at the International Communication Association Meeting, Boston, May 2011.
15. Toynbee, "Fingers to the Bone or Spaced Out on Creativity? Labor Process and Ideology in the Production of Pop."
16. Goldman, *Adventures in the Screen Trade*, 39.
17. Hesmondhalgh, *The Cultural Industries*; Caves, *Creative Industries.*
18. Amman, Carpenter, and Neff, *Surviving the New Economy.*
19. Bernstein, *All Together Now.*
20. Smith and Neuwirth, *The Good Temp*, 64.
21. Kamenetz, *Generation Debt.* For recent news on student debt numbers in the United States, see Scott, "Student Loans Overtake Credit Card Debt."
22. Currid, *The Warhol Economy*; Florida, *The Rise of the Creative Class.*
23. See Benner, *Work in the New Economy*; Batt et al., *Net Working*; and Christopherson, "Project Work in Context."
24. Ross, *No-Collar*, 40.
25. Uchitelle, *The Disposable American.*
26. Bourdieu, *The Field of Cultural Production.*
27. Geertz, *The Interpretation of Cultures.*
28. Durkheim, *Suicide.*
29. Williams, "Business as Usual," 410.
30. Hosein, *Storyteller Uprising.*

Bibliography

Abreu, Elinor. "The Beginning of the End of the End?" *The Standard*. April 4, 2001. http://www.thestandard.com/article/0%2C1902%2C23364%2C00.html (accessed September 3, 2008).

Adkins, Lisa. "The New Economy, Property and Personhood." *Theory, Culture & Society* 22 (1) (2005): 111–130.

Aglietta, Michel, and Régis Breton. "Financial Systems, Corporate Control and Capital Accumulation." *Economy and Society* 30 (4) (2001): 433–466.

Alvear, José, and Derek Top. "Pseudo Bites the Dust." *StreamingMedia.com*. September 18, 2000. http://www.streamingmedia.com/Articles/Editorial/Featured-Articles/Pseudo-Bites-the-Dust-63116.aspx (accessed September 5, 2011).

Amman, John, Tris Carpenter, and Gina Neff. *Surviving the New Economy*. Boulder, Colo.: Paradigm, 2007.

Amoore, Louise. "Risk, Reward and Discipline at Work." *Economy and Society* 33 (2) (2004): 174–196.

Anonymous. "The New York Cyber Sixty." *New York*, November 13, 1995, 44–55, 151.

Anonymous, "Will Content Stay King of New York?" *AtNewYork*, January 1, 1997, www.news-ny.com/news/article.php/250331 (accessed March 17, 2003).

Anonymous. "The VCs Don't Get It and It's Our Own Fault." *AtNewYork*, February 14, 1997.

Anonymous. "1997's Silicon Alley Reporter 100." *Silicon Alley Reporter*, October 1997

Anonymous. "The Map." *Silicon Alley Reporter*, October 1997.

Anonymous. "Web Publishing Ideal Lives as Coalition Funds FEED." *AtNewYork*, December 12, 1997.

Anonymous. "Is Gotham Ready?" *Alley Cat News*, March 2000.

Anonymous. "Net Work News." *Harper's Bazaar*, April 2000.

Anonymous. "Wired." *Harper's Bazaar*, April 2000.

Anonymous, "The Atnewyork.Com Index: Have Alley Stocks Hit Bottom?" *AtNewYork*, April 28, 2000.

Anonymous, "The Breakfast Club," *GQ*, May 2000.

Anonymous. "Xceed Gains Former HBO Vice President." *AtNewYork*, May 3, 2000.

Anonymous. *Silicon Alley Reporter*, December 2000, no. 40.

Anonymous. *Alley Cat News*, February 2001.

Batt, Rosemary, Susan Christopherson, Ned Rightor, and Danielle Van Jaarsveld. *Net Working: Work Patterns and Workforce Policies for the New Media Industry*. Washington, D.C.: Economic Policy Institute, 2001.

Beck, Ulrich. *Risk Society: Toward a New Modernity*. London: Sage, 1992.

Becker, Howard. *Art Worlds*. 25th anniversary ed. Berkeley: University of California Press, 2008.

Benkler, Yochai. *The Wealth of Networks: How Social Production Transforms Markets and Freedom*. New Haven, Conn.: Yale University Press, 2006.

Benner, Chris. *Work in the New Economy*. Oxford: Blackwell, 2002.

Bernstein, Jared. *All Together Now: Common Sense for a Fair Economy*. San Francisco: Berrett-Koehler Publishers, 2006.

Beunza, Daniel, and David Stark. "Tools of the Trade: The Socio-Technology of Arbitrage in a Wall Street Trading Room." *Industrial and Corporate Change* 13 (2) (2004): 369–400.

Black, Sandra E., and Lisa M. Lynch. "What's Driving the New Economy? The Benefits of Workplace Innovation." *Economic Journal* 114 (February 2004): F97–F116.

Blain, Mike, and Gretchen Wilson. "Organizing in the New Economy: The Amazon.com Campaign." *Working USA* 5 (2) (2001): 32–58.

Block, Fred L. *The Vampire State: And Other Myths and Fallacies about the U.S. Economy*. New York: Norton, 1996.

Boltanski, Luc, and Laurent Thévenot. *On Justification: Economies of Worth*. Princeton, N.J.: Princeton University Press, 2006.

Bourdieu, Pierre. *The Field of Cultural Production: Essays on Art and Literature*. Ed. Randal Johnson. New York: Columbia University Press, 1993.

Bowe, Maria. "Word Up! Content's Success Depends on the Right Model." *AtNewYork*, May 8, 1998, http://www.atnewyork.com/news/article.php/250961 (accessed May 12, 2003).

Brenner, Robert. *The Boom and the Bubble: The U.S. in the World Economy*. New York: Verso, 2002.

Brown, Tom. "The '100 Best Companies'; Do They Point the Way for All Business?" Industry Week, April 19, 1993, 12.

Bruegger, Urs, and Karin Knorr Cetina. "Global Microstructures: The Interactional Order of Financial Markets." *American Journal of Sociology* 107 (4) (2002), 905–950.

Burt, Ronald. *Structural Holes: The Social Structure of Competition*. Cambridge, Mass.: Harvard University Press, 1992.

Cantillion, Richard. *Essay on the Nature of Commerce in General*. Trans. and ed. Henry Higgins London: Frank Cass, 1954. http://www.econlib.org/library/NPDBooks/Cantillon/cntNTCover.html (accessed October 7, 2011).

Cappelli, Peter. *The New Deal at Work: Managing the Market-Driven Workforce*. Boston: Harvard Business School Press, 1999.

Castells, Manuel. *The Rise of the Network Society*. Oxford: Blackwell Publishers, 2000.

Caves, Richard. *Creative Industries: Contracts between Art and Commerce*. Cambridge, Mass.: Harvard University Press, 2000.

Centner, Ryan. "Places of Privileged Consumption Practices: Spatial Capital, the Dot-Com Habitus, and San Francisco's Internet Boom." *City & Community* 7 (3) (2008): 193–223.

Chervokas, Jason. "Design Shops Know: Content Is King." *AtNewYork*, January 5, 1996.

Chervokas, Jason. "New York New Media's Ground Zero." *AtNewYork*, April 12, 1996.

Chervokas, Jason. "Online Arts Are More Important to Silicon Alley than Venture Capital" *AtNewYork*, September 19, 1997.

Chervokas, Jason. "In the Evolving Internet Industry, Web Designers Need Not Apply," *AtNewYork*, March 27, 1998.

Chervokas, Jason. "Billions Flowed into Silicon Alley, as Internet Industry Grew Up in 1999." *AtNewYork*, January 6, 2000.

Chervokas, Jason. "Saying Goodbye to Atnewyork.com." *AtNewYork*, May 25, 2000.

Chervokas, Jason, and Tom Watson. Untitled editorial. *AtNewYork*, February 16, 1996.

Chervokas, Jason, and Tom Watson. "The Year the Alley Grew Up." *New York Times*. March 3, 1997. http://partners.nytimes.com/library/cyber/digimet/030397digimet .html (accessed September 26, 2011).

Chervokas, Jason, and Tom Watson. "Everything Old Is New Again." *AtNewYork*, January 5, 2000.

Chet, Carrie Lane. "Like Exporting Baseball: Individualism and Global Competition in the High-Tech Industry." *Anthropology of Work Review* 25 (3–4) (2004): 18–26.

Child, John, and Rita Gunther McGrath. "Organizations Unfettered: Organizational Form in an Information-Intensive Economy." *Academy of Management Journal* 44 (2001): 1135–1148.

Christopherson, Susan. "Project Work in Context: Regulatory Change and the New Geography of Media." *Environment and Planning A* 34 (11) (2002): 2003–2015.

Clark, Gordon L, Nigel Thrift, and Adam Tickell. "Performing Finance: The Industry, the Media and Its Image." *Review of International Political Economy* 11 (2) (2004): 289–310.

Coleman, James S. "Social Capital in the Creation of Human Capital." *The American Journal of Sociology* 94 (1988): S95–S120.

Currid, Elizabeth. *The Warhol Economy: How Fashion, Art, and Music Drive New York City*. Princeton, N.J.: Princeton University Press, 2007.

de Certeau, Michel. *The Practice of Everyday Life*. Trans. Steven F. Rendail. Berkeley: University of California Press, 1984.

De Cock, Christian, James Fitchett, and Christina Volkmann. "Constructing the New Economy: A Discursive Perspective." *British Journal of Management* 16 (2005): 37–49.

Deutsch, Claudia H. "On Electronic Highway, Manhattan Is a Destination." *New York Times*, July 23, 1995, 9:1.

Deuze, Mark. *Media Work*. Cambridge, UK: Polity, 2007.

Douglas, Mary. "Risk as a Forensic Resource." *Daedalus* 119 (4) (1990): 1–16.

Douglas, Mary, and Aaron Wildavsky. *Risk and Culture: An Essay on the Selection of Technological and Environmental Dangers*. Berkeley: University of California Press, 1982.

Drucker, Peter. *Innovation and Entrepreneurship: Practice and Principles*. New York: Harper & Row, 1985.

Duncombe, Stephen. *Notes from Underground: Zines and the Politics of Alternative Culture*. London; New York: Verso, 1997.

Durkheim, Emile. *Suicide*. New York: Free Press, 1951.

Dyson, Esther. "So Many Ideas, So Few Companies." *New York Times Magazine*, February 15, 1998, 16.

Edgecliffe-Johnson, Andrew. "From Dotcom to *Garçon." The Financial Times*, March 14, 2001, 25.

Elias, David. *Dow 40,000.* New York: McGraw-Hill, 1999.

Eliasoph, Nina, and Paul Lichterman. "Culture in Interaction." *American Journal of Sociology* 108 (4) (2003): 735–794.

Flew, Terry. "Creativity, Cultural Studies, and Services Industries." *Communication and Critical/Cultural Studies* 1 (2) (2004): 176–193.

Florida, Richard. *The Rise of the Creative Class: And How It's Transforming Work, Leisure, Community and Everyday Life.* New York: Basic Books, 2004.

Frank, Robert, and Philip J. Cook. *The Winner-Take-All Society: Why the Few at the Top Get So Much More Than the Rest of Us.* New York: Penguin Books, 1996.

Frank, Thomas. *One Market under God: Extreme Capitalism, Market Populism and the End of Economic Democracy.* New York: Doubleday, 2000.

Gabriel, Trip. "Where Silicon Alley Artists Go to Download." *New York Times*, October 8, 1995.

Geertz, Clifford. *The Interpretation of Cultures.* New York: Basic Books, 1973.

Girard, Monique, and David Stark. "Distributing Intelligence and Organizing Diversity in New Media Projects." *Environment and Planning A* 34 (11) (2002): 1927–1949.

Gladwell, Malcolm. "Six Degrees of Lois Weisberg." New Yorker, January 11, 1999, 11.

Gladwell, Malcolm. *The Tipping Point: How Little Things Can Make a Big Difference.* Boston: Little, Brown, 2000.

Glassman, James K., and Kevin A. Hassett. *Dow 36,000.* New York: Times Business, 1999.

Goldman, William. *Adventures in the Screen Trade.* New York: Warner Books, 1983.

Gordon, Robert J. "Hi-Tech Innovation and Productivity Growth: Does Supply Create Its Own Demand?" *National Bureau of Economic Research Working Paper*, no. 9437 (2003).

Grabher, Gernot. "Collective Tinkering: Emerging Project Ecologies in New Media." *Environment and Planning A* 34 (11) (2002): 1911–1926.

Grabher, Gernot. "Ecologies of Creativity: The Village, the Group, and the Heterarchic Organization of the British Advertising Agency." *Environment and Planning A* 33 (2001): 351–374.

Granovetter, Mark S. *Getting a Job: A Study of Contacts and Careers*. Chicago: University of Chicago Press, 1974.

Grigoriadis, Vanessa. "Silicon Alley 10003." *New York Magazine*, March 6, 2000. http://nymag.com/nymetro/news/media/internet/2285/ (accessed September 5, 2011).

Hacker, Jacob S. *The Great Risk Shift*. New York: Oxford University Press, 2006.

Hamm, Steve. "Finding Your Way: Ease of Use on the Internet." *PC Week*, October 24, 1994, A1.

Harmon, Amy. "Beyond Boosterism in the Alley; New York's Web Industry May Be Outgrowing Its Image." *New York Times*, August 17, 1998, D1.

Hesmondhalgh, David. *The Cultural Industries*. 2nd ed. Thousand Oaks, Calif.: Sage, 2007.

Heydebrand, Wolf. Multimedia Networks, Globalization and Strategies of Innovation: The Case of Silicon Alley. In *Multimedia and Regional Economic Restructuring*, ed. Hans-Joachim Braczyk, Gerhard Fuchs, and Hans-Georg Wolf, 40–63. London; New York: Routledge, 1999.

Hirsch, Paul M. "Processing Fads and Fashions: An Organizational-Set Analysis of Cultural Industry Systems." *American Journal of Sociology* 77 (4) (1972): 639–659.

Hochschild, Arlie Russell. *The Managed Heart: Commercialization of Human Feeling*. Berkeley: University of California Press, 1983.

Hodgson, Geoffrey M. *Economics and Utopia: Why the Learning Economy Is Not the End of History*. London: Routledge, 1999.

Hosein, Hanson. *Storyteller Uprising: Trust and Persuasion in the Digital Age*. Seattle: HRH Media, 2011.

Houck, Davis W., and Amos Kiewe, eds. *Actor, Ideologue, Politician: The Public Speeches of Ronald Reagan*. Westport, Conn.: Greenwood Press, 1993.

Howard, Philip. "Network Ethnography and the Hypermedia Organization: New Media, New Organizations, New Methods." *New Media and Society* 4 (4) (2002): 550–574.

IBM. "IBM Highlights: 1990–1995." 2001. http://www-03.ibm.com/ibm/history/documents/pdf/1990-1995.pdf (accessed March 18, 2008).

Indergaard, Michael L. *Silicon Alley: The Rise and Fall of a New Media District*. New York: Routledge, 2004.

Joyce, Erin. "Alley Buzz: Trouble at NYNMA?" *Internet.com*. 2001. http://www.internetnews.com/bus-news/article.php/920871 (accessed June 25, 2009).

Kadlec, Charles W. *Dow 100,000*. New York: NYIF, 1999.

Kadushin, Charles. *American Intellectual Elite*. Boston: Little, Brown, 1974.

Kafka, Peter. "Comcast Buys DailyCandy for $125 Million." August 5, 2008. http://www.alleyinsider.com/2008/8/comcast-buys-dailycandy-for-125-million-beats-out-viacom-for-newsletter-business (accessed August 29, 2008).

Kait, Casey, and Stephen Weiss. *Digital Hustlers: Living Large and Falling Hard in Silicon Alley*. New York: Regan Books, 2001.

Kalleberg, Arne L., Barbara F. Reskin, and Ken Hudson. "Bad Jobs in America: Standard and Nonstandard Employment Relations and Job Quality in United States." *American Sociological Review* 65 (2) (2000): 256–278.

Kamenetz, Anya. *Generation Debt: Why Now Is a Terrible Time to Be Young*. New York: Penguin, 2006.

Kanter, Rosabeth Moss. Careers and the Wealth of Nations: A Macro-Perspective on the Structure and Implications of Career Forms. In *Handbook of Career Theory*, ed. Michael B. Arthur, Douglas T. Hall, and Barbara S. Lawrence, 506–521. Cambridge, UK: Cambridge University Press, 1989.

Kanter, Rosabeth Moss. "Nice Work If You Can Get It: The Software Industry as a Model for Tomorrow's Jobs." *American Prospect* 6 (23) (1995): 52–58.

Kelly, Kevin. "Scan This Book!" *New York Times*, May 14, 2006.

Kelly, Kevin. *What Technology Wants*. New York: Viking, 2010.

Kiewe, Amos, and Davis W. Houck. *A Shining City on a Hill: Ronald Reagan's Economic Rhetoric, 1951–1989*. Praeger Series in Political Communication. New York: Praeger, 1991.

Knight, Frank H. *Risk, Uncertainty, and Profit*. New York: A. M. Kelly, 1964.

Kripner, Greta R. "The Financialization of the American Economy." *Socio-Economic Review* 3 (2005): 173–208.

Kripner, Greta R. "The Financialization of the American Economy." *Socio-Economic Review* 3 (2005): 173–208.

Kunda, Gideon, R. Stephen Barley, and James Evans. "Why Do Contractors Contract? The Experience of Highly Skilled Technical Professionals in a Contingent Labor Market." *Industrial & Labor Relations Review* 55 (2) (2002): 234–261.

Kushner, David. "New York Super Schmooze." November 25, 1997. http://www.wired.com/culture/lifestyle/news/1997/11/8773 (accessed June 25, 2009).

Lash, Scott, and John Urry. *Economies of Signs and Space*. London: Sage, 1994.

Lee, Denny. "A Once-Evocative Name Falls Victim to the Bursting of the High-Tech Bubble." *New York Times*, March 24, 2002, 14:4.

Leonard, Bill. "Networking Still the Best Way to Find Work." *HRMagazine*, December 1, 1999, 28.

Levering, Robert, and Milton Moskowitz. *The 100 Best Companies to Work for in America*. New York: Currency/Doubleday, 1993.

Limerick, Patricia Nelson. "Of Forty-Niners, Oilmen and the Dot-Com Boom." *New York Times*, May 7, 2000, 3:4.

Luhmann, Niklas. *Risk: A Sociological Theory*. New York: A. de Gruyter, 1993.

Lyon, Alexander. "Participants' Use of Cultural Knowledge as Cultural Capital in a Dot-Com Start-Up Organization." *Management Communication Quarterly* 18 (2) (2004): 175–203.

MacKenzie, Donald, Fabian Muniesa, and Lucia Siu. *Do Economists Make Markets?: On the Performativity of Economics*. Princeton, N.J.: Princeton University Press, 2007.

Mandel, Michael J. "The Digital Juggernaut." *Business Week*, Information Revolution Special Issue, May 18, 1994, 22–31.

Mandel, Michael J. "The Job Market Isn't Really in the Dumps." *Business Week*, June 19, 1995, 42.

Mandel, Michael J. "The Triumph of the New Economy." *Business Week*, December 30, 1996, 68.

Mandel, Michael J. *The High-Risk Society: Peril and Promise in the New Economy*. New York: Random House, 1996.

Mandel, Michael J. "You've Got the New Economy All Wrong, Lou." *Business Week*, November 27, 2000, 64.

Manuel, Peter. Cassette Culture: Popular Music and Technology in North India. Chicago: University of Chicago Press, 1993.

Marx, Karl. Estranged Labor. The Economic and Philosophical Manuscripts. Amherst, N.Y.: Prometheus Books, [1844] 1988.

Marx, Karl, and Frederick Engels. *The Communist Manifesto: A Modern Edition*. London; New York: Verso, [1848] 1998.

Massanari, Adrienne. Dot-Coms and Cybercultural Studies: Amazon.com as a Case Study. In *Critical Cybercultural Studies*, ed. David Silver and Adrienne Massanari, 279–293. New York: NYU Press, 2008.

McRobbie, Angela. "From Clubs to Companies: Notes on the Decline of Political Culture in Speeded Up Creative Worlds." *Cultural Studies* 16 (4) (2002): 516–531.

Miles, Matthew B., and A. Michael Huberman. *Qualitative Data Analysis*. 2nd ed. Thousand Oaks, Calif.: Sage, 1994.

Morrissey, Brian. "Primedia Scoops Up gURL.com From Delia's." *Silicon Alley Daily*, May 29, 2001.

Mosco, Vincent. *The Digital Sublime: Myth, Power, and Cyberspace*. Cambridge, Mass.: MIT Press, 2004.

Motavalli, John. *Bamboozled at the Revolution: How Big Media Lost Billions in the Battle for the Internet*. New York: Viking, 2002.

Motta, Katie. "Pace of Dot-Com Layoffs Slows in March." *The Industry Standard*. March 24, 2001. http://www.thestandard.com/pace-dot-com-layoffs-slows-march (accessed August 19, 2008).

Neff, Gina. Associating Independents: Business Relationships of the Dot-Com Era. In *Currents in Critical Cyberstudies*, ed. David Silver and Adrienne Massanari, 294–307. New York: New York University Press, 2006.

Neff, Gina, Elizabeth Wissinger, and Sharon Zukin. "Entrepreneurial Labor among Cultural Producers: 'Cool' Jobs in 'Hot' Industries." *Social Semiotics* 15 (3) (2005): 307–334.

New York New Media Association (NYNMA) and Coopers & Lybrand. "New York New Media Industry Survey: Opportunities and Challenges of New York's Emerging Cyber-Industry." New York: NYNMA, 1996.

New York New Media Association (NYNMA) and Coopers & Lybrand. "2nd New York New Media Industry Survey." New York: NYNMA, 1997.

New York New Media Association (NYNMA) and Price Waterhouse Coopers. "3rd New York New Media Industry Survey: Opportunities and Challenges of NY's Emerging Cyberindustry." New York: NYNMA, 2000.

O'Riain, Sean. "The Flexible Developmental State: Globalization, Information Technology, and the 'Celtic Tiger.'" *Politics & Society* 28 (2) (2000): 157–193.

Pearlstein, Steven. "New Economy Gives Work a Hard Edge." *The Washington Post*, November 14, 1995.

Pearlstein, Steven. "Reshaped Economy Exacts Tough Toll." *The Washington Post*, November 12, 1995, A01.

Peters, Tom. "The Brand Called You." *Fast Company*, no. 10 (1997).

Peters, Tom. "What We Will Do for Work." *Time*, May 2000.

Pham, Alex, and Stephanie Stoughton. "Dot-Com Firms Foster New Corporate Culture: Risk and Reward Are Key, Not Job Loyalty." *The Boston Globe*, August 6, 2000, A26.

Pink, Daniel H. *Free Agent Nation: How America's New Independent Workers Are Transforming the Way We Live*. New York: Warner Business Books, 2002.

Piore, Michael J., and Charles Sabel. *The Second Industrial Divide*. New York: Basic Books, 1984.

Pratt, Andy C. "New Media, the New Economy, and New Spaces." *Geoforum* 31 (2000): 425–436.

Pulitzer, Courtney. "@The Scene." *AtNewYork*, March 21, 1997. http://web.archive.org/web/19980516041614/http://www.atnewyork.com/scene.htm (accessed May 20, 2003).

Pulitzer, Courtney. "@The Scene: Why Silicon Alley Is a Real Community," *AtNewYork*, October 17, 1997.

Pulitzer, Courtney. "@The Scene." *AtNewYork*, November 14, 1997.

Pulitzer, Courtney. "@The Scene." *AtNewYork*, July 11, 1997.

Pulitzer, Courtney. "@The Scene." *AtNewYork*, March 20, 1998.

Pulitzer, Courtney. *The Cyber Scene*, January 29, 1999.

Pulitzer, Courtney. "Razorfishsucks.com." *The Cyber Scene*, May 7, 1999.

Pulitzer, Courtney. "Three Day Rave." *The Cyber Scene*, August 20, 1999.

Pulitzer, Courtney. "Holy Holiday." *The Cyber Scene*, December 22, 2000.

Reich, Robert. "Casualties of the Inflation War." *The Financial Times*, September 24, 1996.

Rose, Nikolas. *Inventing Ourselves: Psychology, Power and Personhood*. Cambridge, UK: Cambridge University Press, 1998.

Ross, Andrew. *No-Collar: The Humane Workplace and Its Hidden Costs*. New York: Basic Books, 2003.

Samuelson, Robert J. "R.I.P.: The Good Corporation." *Newsweek*, July 5, 1993, 41.

Saxenian, AnnaLee. *Regional Advantage: Culture and Competition in Silicon Valley and Route 128*. Cambridge, Mass.: Harvard University Press, 1994.

Schatzman, Leonard, and Anselm Straus. *Field Research Strategies for a Natural Sociology*. Englewood Cliffs, N.J.: Prentice-Hall, 1973.

Scott, Amy. "Student Loans Overtake Credit Card Debt." *Marketplace*, August 10, 2010, http://marketplace.publicradio.org/display/web/2010/08/10/pm-student-loans-overtake-credit-card-debt/ (accessed October 10, 2011).

Sennett, Richard. *The Culture of the New Capitalism*. New Haven, Conn.: Yale University Press, 2006.

Shy, Oz. *The Economics of Network Industries*. Cambridge, UK: Cambridge University Press, 2001.

Sikoryak, Robert. "Hacking It." *New Yorker*, September 25, 1995.

Slater, Don. "From Calculation to Alienation: Disentangling Economic Abstractions." *Economy and Society* 31 (2) (2002): 234–249.

Smith, Mark A. *The Right Talk: How Conservatives Transformed the Great Society into the Economic Society*. Princeton, N.J.: Princeton University Press, 2007.

Smith, Vicki. "New Forms of Work Organization." *Annual Review of Sociology* 23 (1) (1997): 315–339.

Smith, Vicki. *Crossing the Great Divide*. Ithaca, N.Y.: Cornell University Press, 2001.

Smith, Vicki, and Esther B. Neuwirth. *The Good Temp*. Ithaca, N.Y.: Cornell University Press, 2008.

Stark, David. *The Sense of Dissonance: Accounts of Worth in Economic Life*. Princeton, N.J.: Princeton University Press, 2009.

St. John, Warren. "Alive and Well in Silicon Alley." *New York Times*, March 12, 2006.

Stiroh, Kevin. "Is There a New Economy?" *Challenge Magazine* 42 (4) (1999): 82–101.

Terranova, Tiziana. "Free Labor: Producing Culture for the Digital Economy." *Social Text* 18 (2) (2000): 33–58.

Thrift, Nigel. "'It's the Romance Not the Finance That Makes the Business Worth Pursuing': Disclosing a New Market Culture." *Economy and Society* 30 (4) (2001): 412–432.

Toynbee, Jason. Fingers to the Bone or Spaced Out on Creativity? Labor Process and Ideology in the Production of Pop. In *Cultural Work: Understanding Cultural Industries*, ed. Andrew Beck, 39–54. London: Routledge, 2003.

Tractenberg, Jeffrey A. "New York Grows as a Multimedia Hub as More Firms Set Up Projects There." *Wall Street Journal*, August 24, 1994.

Turner, Fred. *From Counterculture to Cyberculture: Stewart Brand, the Whole Earth Network, and the Rise of Digital Utopianism*. Chicago: University of Chicago Press, 2006.

Uchitelle, Louis. *The Disposable American: Layoffs and Their Consequences*. New York: Knopf, 2006.

U.S. Department of Labor, Bureau of Labor Statistics. *Occupational Outlook Handbook 2000–2001*. Lincolnwood, Ill.: VGM Career Horizons, 2001.

Van Loon, Joost. *Risk and Technological Culture: Towards a Sociology of Virulence.* London: Routledge, 2002.

Van Slambrouck, Paul. "America's Growing Affinity for Risk." *Christian Science Monitor*, December 28, 1999.

Washington Alliance of Technology Workers. "The State of Seattle Area IT Employment and Training: Results of the IT Employer and IT Employee Surveys." Seattle, Wash., 2002.

Watson, Tom. "What the Industry Needs: Real Representation from NYNMA." *AtNewYork*, December 19, 1997.

Watson, Tom. "Where Content Is King: Portrait of a Figurehead." *AtNewYork*, March 13, 1998.

Weil, Benjamin. "Can Consumer Companies Profit from Web Art Experiments?" *AtNewYork*, February 6, 1998.

Weinstein, Bob. "Make Networking a Career Habit." *Chicago Sun-Times*, May 17, 1998.

White, Harrison C. *Markets from Networks: Socioeconomic Models of Production.* Princeton, N.J.: Princeton University Press, 2001.

Whyte, William H. *The Organization Man.* New York: Doubleday, 1956.

Williams, Karel. "Business as Usual." *Economy and Society* 30 (4) (2001): 399–411.

Zook, Matthew A. "The Web of Production: The Economic Geography of Commercial Internet Content Production in the United States." *Environment and Planning A* 32 (2000): 411–426.

Index